SPRINGER SERIES IN PHOTONICS 5

Springer
Berlin
Heidelberg
New York
Barcelona
Hong Kong
London
Milan
Paris
Tokyo

Physics and Astronomy

ONLINE LIBRARY

http://www.springer.de/phys/

SPRINGER SERIES IN PHOTONICS

Series Editors: T. Kamiya   B. Monemar   H. Venghaus   Y. Yamamoto

The Springer Series in Photonics covers the entire field of photonics, including theory, experiment, and the technology of photonic devices. The books published in this series give a careful survey of the state-of-the-art in photonic science and technology for all the relevant classes of active and passive photonic components and materials. This series will appeal to researchers, engineers, and advanced students.

J. Kim   S. Somani   Y. Yamamoto

Nonclassical Light from Semiconductor Lasers and LEDs

With 113 Figures

 Springer

Dr. Jungsang Kim
Bell Laboratories
Lucent Technologies
600 Mountain Avenue,
Murray Hill, NJ 07974, USA

Professor Yoshihisa Yamamoto
Stanford University
Edward L. Ginzton Laboratory
Stanford, CA 94305, USA

Dr. Seema Somani
Schlumberger Technologies
150 Baytech Drive
San Jose, CA 95134, USA

Series Editors:

Professor Takeshi Kamiya
Dept. of Electronic Engineering
Faculty of Engineering
University of Tokyo
7-3-1 Hongo, Bunkyo-ku, Tokyo, 113, Japan

Dr. Herbert Venghaus
Heinrich-Hertz-Institut
für Nachrichtentechnik Berlin GmbH
Einsteinufer 37
10587 Berlin, Germany

Professor Bo Monemar
Department of Physics
and Measurement Technology
Materials Science Division
Linköping University
58183 Linköping, Sweden

Professor Yoshihisa Yamamoto
Stanford University
Edward L. Ginzton Laboratory
Stanford, CA 94305, USA

ISSN 1437-0379
ISBN 3-540-67717-8 Springer-Verlag Berlin Heidelberg New York

Library of Congress Cataloging-in-Publication Data: Kim, J. (Jungsang), 1969– . Nonclassical light from semiconductor lasers and LEDs/ J. Kim, S. Somani, Y. Yamamoto. p. cm. – (Springer series in photonics, ISSN 1437-0379; 5). Includes bibliographical references and index. ISBN 3540677178 (alk. paper) 1. Squeezed light. 2. Semiconductor lasers. 3. Light emitting diodes. I. Somani, S. (Seema), 1970– II. Yamamoto, Yoshihisa. III. Title. IV. Springer series in photonics; v. 5. QC446.3.S67 K56 2001   535–dc21   2001031178

Springer-Verlag Berlin Heidelberg New York
a member of BertelsmannSpringer Science+Business Media GmbH

http://www.springer.de

© Springer-Verlag Berlin Heidelberg 2001
Printed in Germany

Data conversion: Frank Herweg, Leutershausen
Cover concept: eStudio Calamar Steinen
Cover production: *design & production* GmbH, Heidelberg

Printed on acid-free paper   SPIN: 10771938   57/3141/ba   5 4 3 2 1 0

Preface

The quantum statistical properties of light generated in a semiconductor laser and a light-emitting diode (LED) have been a field of intense research for more than a decade. This research monograph discusses recent research activities in nonclassical light generation based on semiconductor devices, performed mostly at Stanford University.

When a semiconductor material is used as the active medium to generate photons, as in semiconductor lasers and LEDs, the flow of carriers (electrons and holes) is converted into a flow of photons. Provided that the conversion is fast and efficient, the statistical properties of the carriers ("pump noise") can be transferred to the photons; if pump noise can be suppressed to below the shot noise value, the noise in the photon output can also be suppressed below the Poisson limit. Since electrons and holes are fermions and have charges, the statistical properties of these particles can be significantly different from those of photons if the structure of the light-emitting device is properly designed to provide interaction between these particles.

There has been a discrepancy between the theoretical understanding and experimental observation of noise in a macroscopic resistor until very recently. The dissipation that electrons experience in a resistor is expected to accompany the fluctuation due to partition noise, leading to shot noise in the large dissipation limit as is the case with photons. Experimental observation shows that thermal noise, expected only in a thermal-equilibrium situation (zero-bias condition), is the only source of noise featured by a resistor, independent of the current. Justification for using thermal-noise formulae for the noise power spectral density in a resistor in the presence of a finite current (nonequilibrium situation) requires a microscopic understanding of the dissipation process in electron systems. Suppression of current shot noise (generation of "quiet electron flows") in a macroscopic resistor has been discussed recently in the context of mesoscopic physics. It was found that the Pauli exclusion principle between electrons is ultimately responsible for the absence of shot noise in a dissipative resistor. This shot noise suppression by inelastic electron scattering in a source resistor guarantees that the current supplied to the junction by the external circuit carries only thermal noise and is often interpreted as the direct physical origin for pump-noise suppression, leading to the generation of "quiet photon flows (intensity-squeezed light)".

It has been pointed out recently that the carriers supplied by the external circuit must be injected stochastically across the depletion layer of the p–n junction. This means that the output photons can carry full shot noise even if the external circuit current noise is completely suppressed, as long as the stochasticity in the carrier injection process is not regulated. The charging energy at the junction plays a key role in establishing the correlation between successive carrier-injection events, thereby eliminating the stochasticity.

A laser is an extremely nonlinear device in which the optical gain of a specific mode strongly depends on the optical feedback of the mode. Such nonlinearity can potentially enhance the noise of a single output mode, especially in a semiconductor laser, where several cavity modes compete for optical gain from a spectrally broad gain medium. Conversion of the quiet stream of carriers into photons is not enough to guarantee a quiet photon stream in a single output mode of a semiconductor laser. Noise associated with the competition of multiple modes (mode partition noise) need to be considered, and a laser structure that minimizes the mode partition noise needs to be constructed. Once all these issues are resolved, one ends up with a semiconductor laser that produces sub-shot-noise output in the lasing mode. The quiet stream of photons generated in this way can be potentially useful for high-precision measurements such as laser spectroscopy and interferometry.

In the squeezing experiments with a macroscopic p–n junction, however, only a large number of photons on the order of 10^8 can be regulated. An interesting question is whether such a regulation mechanism can be extended all the way down to the single photon level.

When the single-electron charging energy (e^2/C, where C is the capacitance associated with the tunnel junction) of an ultra-small tunnel junction is larger than the thermal energy ($k_B T$), the tunneling of electrons can be mutually correlated due to the Coulomb repulsive interaction between successive electrons. The first theoretical prediction and experimental demonstration of this effect were reported in the late 1980s and became widely known as the Coulomb blockade effect. Simultaneous Coulomb blockade effects for electrons and holes can exist in a mesoscopic p–n junction, and nonclassical light can be generated from such a structure. A single-photon turnstile device is herein proposed, where a single electron and a single hole are injected into the optically active region of a p–n junction to generate a stream of regulated single photons.

Efficient detection of such single-photon states is another challenge. Several technical breakthroughs have enabled high-efficiency and low-noise single-photon detection. Among the single photon-detectors developed so far, visible-light photon counters (VLPCs) have demonstrated the highest quantum efficiency and lowest multiplication noise properties. This particular detector was used to demonstrate single-photon turnstile device operation, and promises advances in quantum optics experiments using single photons and entangled photon pairs.

This book is composed of four major parts. The first part discusses the generation of sub-shot-noise light in macroscopic p–n junction light-emitting devices, including semiconductor laser and LEDs. The second part discusses the application of squeezed light to high-precision measurements, including spectroscopy and interferometry. The third part considers the Coulomb blockade effect in a mesoscopic p–n junction and the generation of single-photon states, and the last part addresses the detection of single photons using a VLPC.

Chapter 1 gives an overview of the classical and quantum descriptions of an electromagnetic field. In Chap. 2, the mechanism for suppression of carrier injection noise across the depletion layer of a macroscopic p–n junction is discussed. Such noise suppression is responsible for generation of amplitude-squeezed states in a semiconductor laser and sub-Poissonian light in an LED. Chapter 3 describes an experiment that provides experimental evidence of such a noise suppression mechanism at work. Chapter 4 discusses the measurement of longitudinal-mode-partition noise in a semiconductor laser. Careful analysis of the laser structure indicates that the negative correlation responsible for perfect cancellation of longitudinal mode partition noise can be destroyed by saturable absorbers. A model describing this mechanism is presented in Chap. 5. Based on these studies, a careful procedure was developed to fabricate laser structures with low saturable absorption. Consistent squeezing was observed from those lasers, and the details of the development are outlined in Chap. 6. These experiments were performed in collaboration with Dr. Hirofumi Kan at Hamamatsu Photonics in Japan.

In Chap. 7, sub-shot-noise frequency modulation (FM) spectroscopy of cold cesium atoms is discussed. Chapter 8 describes sub-shot-noise FM noise spectroscopy and phase-sensitive noise spectroscopy using semiconductor lasers. These experiments were done in collaboration with Dr. Steven Kasapi. An experiment on sub-shot-noise interferometers is discussed in Chap. 9, performed in collaboration with Dr. Shuichiro Inoue.

In Chaps. 10 and 11, detailed studies of a mesoscopic p–n junction light-emitting device are presented. The discussions are focused towards the realization of a single-photon turnstile device based on the principle of Coulomb blockade. In Chap. 10, the operation principle for a single-photon turnstile device is presented which is based on parameters that are achievable in a real experimental situation. Monte Carlo simulations of this device were performed in collaboration with Dr. Oliver Benson. Chapter 11 describes the experimental effort towards the realization of a sub-micron LED in the GaAs material system where the Coulomb blockade effect is observed.

In Chap. 12, the characterization of a single-photon counting detector based on a VLPC is presented. The series of experiments demonstrates the highest single-photon detection quantum efficiency, the noise-free avalanche multiplication process, and the multi-photon detection capability of this unique detector. The detectors were provided by Dr. Henry H. Hogue of Boeing North American, and the measurements were performed in collaboration with Dr. Shigeki Takeuchi of Mitsubishi Electric Company.

In Chap. 13, future directions for nonclassical light generation in semiconductor devices are discussed. Light sources based on single quantum dots and microcavities are the main topic of this chapter. Generation of regulated single photons and entangled photon-pairs using a single quantum dot microcavity system will be discussed, as well as nonclassical matter-wave generation using stimulated exciton-exciton scattering in a quantum sell microcavity system. These works were performed in collaboration with Dr. Glenn Solomon, Dr. Francesco Tassone, Robin Huang, Matthew Pelton and Charles Santori.

We would like to thank our colleagues for helpful collaboration and valuable discussions, in particular, Hirofumi Kan, Kazunori Tanaka, Susumu Machida, Shuichiro Inoue, Steven Kasapi, Oliver Benson, Henry H. Hogue, Shigeki Takeuchi, Glenn Solomon, Francesco Tassone, Robin Huang, Matthew Pelton and Charles Santori. Finally, we wish to express our thanks to Mayumi Hakkaku for her efficient word processing.

Murray Hill, San Jose, and Stanford   *Jungsang Kim*
February 2001   *Seema Somani*
Yoshihisa Yamamoto

Contents

1. Nonclassical Light

1.1 Classical Description of Light

A classical single-mode electromagnetic field is completely determined by two real numbers, amplitude and phase:

$$\begin{aligned}
E(t) &= E \cos(\omega t - \phi)u(r) \\
&= (a_1 \cos \omega t + a_2 \sin \omega t)u(r) \\
&= \frac{1}{2}(a\, e^{-i\omega t} + a^*\, e^{i\omega t})u(r) \quad,
\end{aligned} \tag{1.1}$$

where $a_1 = E \cos \phi = \frac{1}{2}(a + a^*)$ and $a_2 = E \sin \phi = \frac{1}{2i}(a - a^*)$ are the two quadrature amplitudes and ϕ is the phase. The electric field amplitude E is related to the photon number n by

$$E = \mathcal{E}_0 \sqrt{n} \quad, \tag{1.2}$$

where

$$\mathcal{E}_0 = \sqrt{\frac{\hbar\omega}{\varepsilon_0 V}} \quad, \tag{1.3}$$

is the electric field created by one photon in a mode volume V. The spatial mode function $u(r)$ is determined by the boundary condition and satisfies the normalization

$$\int |u(r)|^2 dr^3 = V \quad. \tag{1.4}$$

At a finite temperature, the thermal background noise must be added to the two quadrature amplitudes a_1 and a_2 with identical (phase independent) noise powers of

$$\begin{aligned}
\langle \Delta a_1^2 \rangle = \langle \Delta a_2^2 \rangle &= \frac{\left(\frac{1}{2}\right)}{e^{\hbar\omega/k_B T} - 1} \\
&\simeq \frac{1}{2}\left(\frac{k_B T}{\hbar\omega}\right) \quad.
\end{aligned} \tag{1.5}$$

Here the last equality is based on the assumption that $k_B T \gg \hbar\omega$ (classical limit). The noise distribution is isotropic and independent of the phase as shown in Fig. 1.1.

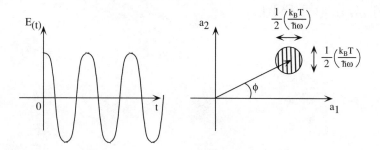

Fig. 1.1. A classical electromagnetic field with thermal background noise

1.2 Quantum Description of Light

The classical description of an electromagnetic field, given in Sect. 1.1, is valid
if the thermal noise is much greater than the quantum mechanical zero-point
fluctuation ($k_B T \gg \hbar\omega/2$). In the spectral ranges of infrared, visible, ultraviolet and X-ray radiation, this condition is violated even at room temperature,
and we must take the quantum description of light.

Canonical quantization of an electromagnetic field is achieved by replacing
real number amplitudes a_1 and a_2 by Hilbert-space Hermitian operators $\hat{a}_1$
and $\hat{a}_2$ and imposing the commutation relation

$$[\hat{a}_1, \hat{a}_2] = \frac{i}{2} \quad , \tag{1.6}$$

or equivalently

$$[\hat{a}, \hat{a}^\dagger] = 1 \quad , \tag{1.7}$$

where $\hat{a} = \hat{a}_1 + i\hat{a}_2$ and $\hat{a}^\dagger = \hat{a}_1 - i\hat{a}_2$ are the photon creation and annihilation
operators, respectively. Using standard algebra [55, 160] for (1.6), we can
easily derive the uncertainty product of $\hat{a}_1$ and $\hat{a}_2$:

$$\langle \Delta\hat{a}_1^2 \rangle \langle \Delta\hat{a}_2^2 \rangle \geq \frac{1}{16} \quad . \tag{1.8}$$

The minimum uncertainty states of light, which satisfy the equality in (1.8),
are called coherent states [76] and squeezed states [218, 228, 274]. We will
discuss the properties of such quantum states of light in Sect. 1.3.

In contrast to the canonical quantization of the two quadrature amplitudes a_1 and a_2, the quantization of the photon number n and the phase ϕ
requires more careful consideration. The difficulty stems from the absence of
the Hermitian phase operator, which is a direct analog to the classical phase
ϕ in (1.1) [55, 220]. The most successful approach so far [187] defines the
phase operator in the (truncated) $(s + 1)$-dimensional Hilbert space:

$$\hat{\phi}_s = \sum_{m=0}^{s} \theta_m |\theta_m\rangle\langle\theta_m| \quad , \tag{1.9}$$

where $|\theta_m\rangle$ is a complete set of orthonormal phase eigenstates:

$$|\theta_0\rangle = \frac{1}{\sqrt{s+1}} \sum_{n=0}^{s} e^{in\theta_0} |n\rangle \quad , \tag{1.10}$$

$$|\theta_m\rangle = \exp\left[i\left(\frac{2\pi}{s+1}m\right)\hat{n}\right]|\theta_0\rangle, \quad m = 0, 1, \cdots s \quad . \tag{1.11}$$

The distinct phase states can occur for values $\theta_m = \theta_0 + 2\pi m/(s+1)$, and at the maximum extent of the photon number, $s \to \infty$, the phase eigenvalues become a countable infinite set that has a one-to-one correspondence to the photon number eigenvalues. The commutation relation for the photon number and phase operators is given by [187]

$$\langle[\hat{\phi}_s, \hat{n}]\rangle = -i[1 - 2\pi P(\theta_0)] \quad . \tag{1.12}$$

Here $P(\theta_0)$ is the probability of finding the phase of an electromagnetic field at θ_0. In the limit of large s, $P(\theta_0)$ is much smaller than $1/2\pi$ unless θ_0 is chosen at the peak of the phase distribution $P(\theta)$ and, in this case, the uncertainty relation for $\hat{n}$ and $\hat{\phi}_s$ is

$$\langle\Delta\hat{n}^2\rangle\langle\Delta\hat{\phi}_s^2\rangle \geq \frac{1}{4} \quad . \tag{1.13}$$

The minimum uncertainty states of light which satisfy the equality in (1.13) are called number-phase squeezed states [111, 127], which will also be discussed in Sect. 1.3.

1.3 Coherent State, Squeezed State and Number-Phase Squeezed State

The minimum uncertainty state for the two quadrature amplitudes is mathematically defined as the eigenstate of the non-Hermitian operator $e^r \hat{a}_1 + i e^{-r} \hat{a}_2$:

$$e^r(\hat{a}_1 - \langle\hat{a}_1\rangle)|\psi\rangle = -i e^{-r}(\hat{a}_2 - \langle\hat{a}_2\rangle)|\psi\rangle \quad . \tag{1.14}$$

The uncertainty product of the thus defined state can easily be evaluated by (1.14) and satisfies the minimum value

$$\begin{array}{l} \langle\Delta\hat{a}_1^2\rangle = \frac{1}{4}e^{-2r} \\ \langle\Delta\hat{a}_2^2\rangle = \frac{1}{4}e^{2r} \end{array} \Big\rangle \langle\Delta\hat{a}_1^2\rangle\langle\Delta\hat{a}_2^2\rangle = \frac{1}{16} \quad . \tag{1.15}$$

The minimum uncertainty state the special case $r = 0$ is the eigenstate of the photon annihilation operator $\hat{a}$ and is termed a coherent state [76].

The uncertainties of the two quadrature amplitudes are equal— $\langle\Delta\hat{a}_1^2\rangle = \langle\Delta\hat{a}_2^2\rangle = \frac{1}{4}$—for a coherent state. In the general case of $r \neq 0$, the minimum uncertainty state is the eigenstate of the Bogoliubov transformation of the photon creation and annihilation operators, $\hat{b} = \mu\hat{a} + \nu\hat{a}^{\dagger}$, where $\mu = \cosh(r)$ and $\nu = \sinh(r)$, and is termed a squeezed state [228]. The two quadrature amplitudes have different uncertainties but satisfy the minimum uncertainty product.

The minimum uncertainty state for the photon number and the phase is defined as the eigenstate of the non-Hermitian operator $e^r\hat{n} + i\,e^{-r}\hat{\phi}$ [111]:

$$e^r(\hat{n} - \langle\hat{n}\rangle)|\psi\rangle = -i\,e^{-r}(\hat{\phi} - \langle\hat{\phi}\rangle)|\psi\rangle \quad . \tag{1.16}$$

The uncertainty product of the thus defined state satisfies the minimum value

$$\begin{matrix}\langle\Delta\hat{n}^2\rangle = \frac{1}{2}e^{-2r} \\ \langle\Delta\hat{\phi}^2\rangle = \frac{1}{2}e^{2r}\end{matrix} \searrow \atop \nearrow \langle\Delta\hat{n}^2\rangle\langle\Delta\hat{\phi}^2\rangle = \frac{1}{4} \quad . \tag{1.17}$$

In the limits $r \to +\infty$ and $r \to -\infty$, the states approach the photon number eigenstate $|n\rangle$ and the phase eigenstate $|\phi\rangle$.

In Fig. 1.2, the (heuristic) noise distributions of the various quantum states of light are plotted in the $a_1 - a_2$ phase space.

1.4 Quantum Theory of Photodetection and Sub-Poisson Photon Distribution

One of the important detection schemes, which reveals the quantum mechanical nature of light, is photon-counting detection. The quantum mechanical photon-count distribution is given by [119]

$$P_n(T) = \mathrm{Tr}\left(\hat{\rho}\mathcal{N}\frac{(\eta T\hat{I})^n}{n!}\exp(-\eta T\hat{I})\right) \quad , \tag{1.18}$$

where $\hat{\rho}$ and $\hat{I}$ are the field-density operator and the time-averaged intensity operator, respectively. $\mathcal{N}$ is the normal ordering operator, η is a constant proportional to the detector quantum efficiency and T is the detection time interval. In the case of a single-mode field,

$$\hat{I} = \frac{2\varepsilon_0 c}{T}\int_0^T \hat{E}^{(-)}(r,t)\hat{E}^{(+)}(r,t)\mathrm{d}t = \frac{\hbar\omega c}{L^3}\hat{a}^{\dagger}\hat{a} \quad , \tag{1.19}$$

where $\hat{E}^{(-)}(\propto \hat{a}^{\dagger})$ and $\hat{E}^{(+)}(\propto \hat{a})$ are the negative and positive frequency components of the electric field operator, $\hat{E} = \hat{E}^{(-)} + \hat{E}^{(+)}$, and L^3 is a quantization volume. If we define the detector quantum efficiency by $\zeta = \hbar\omega cT\eta/L^3$, (1.18) can be rewritten as

$$P_n(T) = \mathrm{Tr}\left(\hat{\rho}\mathcal{N}\frac{(\zeta\hat{a}^{\dagger}\hat{a})^n}{n!}\exp[-\zeta\hat{a}^{\dagger}\hat{a}]\right) \quad . \tag{1.20}$$

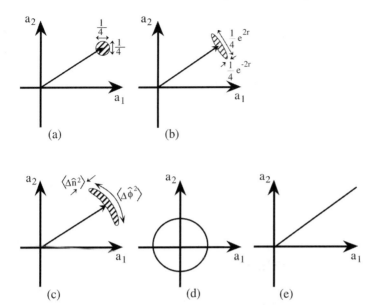

Fig. 1.2 a–e. The noise distributions of various quantum states. (**a**) Coherent state with $\langle \Delta \hat{a}_1^2 \rangle = \langle \Delta \hat{a}_2^2 \rangle = \frac{1}{4}$, (**b**) squeezed state with $\langle \Delta \hat{a}_1^2 \rangle = \frac{1}{4} e^{-2r}$ and $\langle \Delta \hat{a}_2^2 \rangle = \frac{1}{4} e^{2r}$, (**c**) number-phase squeezed state with $\langle \Delta \hat{n}^2 \rangle = \frac{1}{2} e^{-2r}$ and $\langle \Delta \hat{\phi}^2 \rangle = \frac{1}{2} e^{2r}$, (**d**) photon number eigenstate with $\langle \Delta \hat{n}^2 \rangle = 0$, and (**e**) phase eigenstate with $\langle \Delta \hat{\phi}^2 \rangle = 0$. The plots are heuristic noise distributions and do not correspond to any mathematical distribution function

We can evaluate the expression in (1.20) by using the Taylor series expansion of the exponential. Using the diagonal elements P_i of the field-density operator, we obtain

$$
\begin{aligned}
P_n(T) &= \sum_{i=0}^{\infty} P_i \frac{\zeta^n}{n!} \langle i | \sum_{m=0}^{\infty} (-1)^m \frac{\zeta^m}{m!} (\hat{a}^\dagger)^{n+m} (\hat{a})^{n+m} | i \rangle \\
&= \sum_{i=n}^{\infty} P_i \frac{\zeta^n}{n!} \sum_{m=0}^{i-n} \frac{\zeta^m}{m!} \frac{i!}{(i-n-m)!} \\
&= \sum_{i=n}^{\infty} P_i \frac{i!}{n!(i-n)!} \zeta^n (1-\zeta)^{i-n} \quad .
\end{aligned}
\tag{1.21}
$$

To obtain the last equality, we used the fact that summation over m in the second equality corresponds to a binomial distribution. The final expression in (1.21) means that photodetection is a process of random deletion of photons with the probability $(1-\zeta)$ for the initial photon distribution P_i of the light field.

For coherent states, the photon-number distribution is Poissonian, that is, $P_i = \langle \hat{n} \rangle^i e^{-\langle \hat{n} \rangle}/i!$, where $\langle \hat{n} \rangle$ is the average photon number. The photon-counting statistics maintain the Poisson distribution with the average number of $\zeta \langle n \rangle$ after random deletion of photons. The statistical mixture of coherent states always features the super-Poisson photon distribution.

The number-phase squeezed state defined by (1.16) with $r > \ln(1/\sqrt{2\langle \hat{n} \rangle})$ and the photon-number state, however, shows a sub-Poissonian photon distribution. A sub-Poissonian photon-count distribution cannot be described by classical photodetection theory [159, 164, 241], so those states with sub-Poissonian photon distributions are called nonclassical light. According to the classical theory of photodetection, photon-counting statistics obey the Poisson distribution even if the intensity of the field is completely constant.

1.5 Quantum Theory of Second-Order Coherence and Photon Antibunching

Another important detection scheme is the measurement of intensity correlation, often referred to as the Hanbury Brown and Twiss experiment [87]. In this experiment, an incoming field $\hat{E}(r, t)$ is split into two using a 50%–50% beam splitter, and the split beams are detected by two photodetectors located at the different positions r_1 and r_2 as shown in Fig. 1.3. The joint photon-counting rate at the two detectors is proportional to the second-order coherence function [76]:

$$G^{(2)}(\tau) = \langle \hat{E}^{(-)}(t)\hat{E}^{(-)}(t+\tau)\hat{E}^{(+)}(t+\tau)\hat{E}^{(+)}(t) \rangle \quad . \tag{1.22}$$

The normalized second-order coherence function directly follows from (1.22):

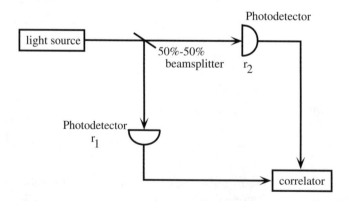

Fig. 1.3. A Hanbury Brown–Twiss experiment for measuring the second order coherence function

$$g^{(2)}(\tau) = \frac{\langle \hat{a}^\dagger(t)\hat{a}^\dagger(t+\tau)\hat{a}(t+\tau)\hat{a}(t)\rangle}{\langle \hat{a}^\dagger(t)\hat{a}(t)\rangle\langle \hat{a}^\dagger(t+\tau)\hat{a}(t+\tau)\rangle} \quad . \tag{1.23}$$

We evaluate $g^{(2)}(\tau)$ for particular quantum states of light. If the state of the field has a Lorentzian distribution of the frequencies ω and each frequency component is in a thermal state, then

$$
\begin{aligned}
\hat{\rho}_\omega &= \left[1 - \exp\left(-\frac{\hbar\omega}{k_B\theta}\right)\right]\exp\left(-\frac{\hbar\omega\hat{a}^\dagger\hat{a}}{k_B\theta}\right) \\
&= \left[1 - \exp\left(-\frac{\hbar\omega}{k_B\theta}\right)\right]\sum_n \exp\left(-\frac{n\hbar\omega}{k_B\theta}\right)|n\rangle\langle n| \quad ,
\end{aligned}
\tag{1.24}
$$

where θ is the temperature of the radiation source. Equation (1.23) can be evaluated for this multi-mode chaotic light:

$$
\begin{aligned}
g^{(2)}(\tau) &= 1 + |g^{(1)}(\tau)|^2 \\
&= 1 + \exp(-\gamma|\tau|) \quad .
\end{aligned}
\tag{1.25}
$$

Here $g^{(1)}(\tau) = \langle \hat{a}^\dagger(t+\tau)\hat{a}(t)\rangle/\sqrt{\langle \hat{a}^\dagger(t+\tau)\hat{a}(t+\tau)\rangle\langle \hat{a}^\dagger(t)\hat{a}(t)\rangle}$ is the first-order coherence function [76] and γ is the linewidth of a Lorentzian spectrum. For thermal light, the detection of a photon makes a second subsequent detection event very likely within the field coherence time γ^{-1}. This phenomenon is referred to as photon bunching.

Next, we consider a single-mode coherent state. The evaluation of $g^{(2)}(\tau)$ is particularly easy for this case, since the coherent state $|\alpha\rangle$ is the eigenstate of the annihilation operator $\hat{a}$. In general, one can show that

$$g^{(n)}(t_1 \cdots t_n) = \frac{\langle \hat{a}^\dagger(t_1)\cdots\hat{a}^\dagger(t_n)\hat{a}(t_n)\cdots\hat{a}(t_1)\rangle}{\langle \hat{a}^\dagger(t)\hat{a}(t)\rangle^n} = 1 \quad . \tag{1.26}$$

The above equation tells us that the coherent state is coherent in all orders. Photon detection events for a coherent state are completely uncorrelated; detection of a photon at $\tau = 0$ gives us no information regarding the detection time of a second photon.

For a single-mode photon number state $|n\rangle$, we obtain

$$g^{(2)}(0) = \frac{n-1}{n} < 1 \quad . \tag{1.27}$$

The above relation tells us that the detection of a photon makes a second subsequent detection event less likely; the phenomenon is known as photon antibunching. It is easy to understand the origin of photon antibunching in a photon-number eigenstate: the total number of photons is known and fixed, and so the detection of one makes it less likely to detect other photons. Particularly, for a single-photon state $|1\rangle$, $g^{(2)}(0) = 0$.

Using simple operator algebra, we may show that [159]

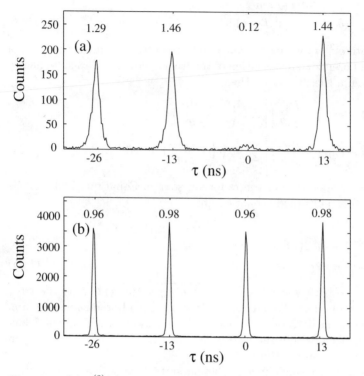

Fig. 1.4. (a) $g^{(2)}(\tau)$ measured for photons emitted from a single quantum dot optically excited by a pulsed Ti:Al$_2$O$_3$ laser, and (b) $g^{(2)}(\tau)$ measured for the pump Ti:Al$_2$O$_3$ laser

$$g^{(2)}(0) = 1 + \frac{\langle \Delta \hat{n}^2 \rangle - \langle \hat{n} \rangle}{\langle \hat{n} \rangle^2} \quad . \tag{1.28}$$

We therefore see that photon antibunching for a single-mode field implies sub-Poisson statistics $\langle \Delta \hat{n}^2 \rangle < \langle \hat{n} \rangle$. If we take the classical description of light, we can use Cauchy's inequality to show that

$$g^{(2)}_{\text{classical}}(0) \geq 1 \quad , \tag{1.29}$$

$$g^{(2)}_{\text{classical}}(0) \geq g^{(2)}_{\text{classical}}(\tau) \quad . \tag{1.30}$$

Thus, photon antibunching, $0 \leq g^{(2)}(0) < 1$, is the other signature of non-classical light.

Strong signatures of photon antibunching can be observed from photons emitted from a single two-level system, where the medium is saturated by a single excitation. Examples of such two-level system include single atoms or ions [53], single molecules [14, 51] or single quantum dots [203]. Figure 1.4a shows an example of the photon antibunching observed for light from a single quantum dot. When a single quantum dot is optically excited by a mode-locked Ti:Al$_2$O$_3$ laser, one and only one electron-hole pair is created and only one photon is emitted per pulse from the dot. Therefore, if one photon is detected at $\tau = 0$, there is no probability of detecting another photon within the same pulse. That is, $g^{(2)}(0) = 0$ for a single photon state. For comparison, the measured $g^{(2)}(\tau)$ for the pump Ti:Al$_2$O$_3$ laser pulse is shown in Figure 1.4(b). Since the Ti:Al$_2$O$_3$ laser pulse is in a coherent state, the detection of one photon at $\tau = 0$ does not influence the subsequent photon detection event at all. The equal heights of the multipeaks in Fig. 1.4b clearly shows the unique behavior of a coherent state.

Hanbury Brown and Twiss experiment has been generalized recently to characterize the quantum statistical properties of various particle sources [94, 183].

1.6 Quantum Theory of Photocurrent Fluctuation and Squeezing

The third important detection scheme, which reveals the quantum mechanical nature of light, is the measurement of a photocurrent fluctuation spectrum. The photocurrent power spectrum is given by the Fourier transform of the two-time correlation function of the photocurrent:

$$S(\omega) = \frac{1}{\pi} \int_0^\infty \mathrm{d}\tau \cos(\omega\tau) \; \overline{i(0)i(\tau)} \quad . \tag{1.31}$$

The quantum mechanical photocurrent correlation function is expressed by [159]

$$\overline{i(0)i(\tau)} = e^2\zeta^2 \langle \hat{a}^\dagger(0)\hat{a}^\dagger(\tau)\hat{a}(\tau)\hat{a}(0)\rangle + e^2\zeta\langle \hat{a}^\dagger(0)\hat{a}(0)\rangle\delta(t) \quad , \tag{1.32}$$

where ζ is the detector quantum efficiency. The second term in this expression is the shot-noise contribution to the current. The first term is the intensity fluctuation, which is zero for a coherent state and positive definite for a statistical mixture of coherent states.

The normally ordered two-time correlation function, the first term of (1.32), can be negative for such quantum states of light as a number-phase squeezed state and photon-number state. Therefore, the sub-shot-noise photocurrent power spectrum is another signature of nonclassical light.

A typical experimental setup for measuring the photocurrent power spectrum is shown in Fig. 1.5. The test field is divided by a 50%–50% beam splitter and converted to photocurrents by two photodetectors. One of the two photodetector output currents is delayed by τ. The difference in the two photocurrents is produced by a differential amplifier. At a fluctuation frequency Ω_in satisfying $\Omega_\mathrm{in}\tau = 2N\pi$ (N: integer), the differential amplifier produces the difference signal, $i_1 - i_2$, of the two photodetector outputs. In

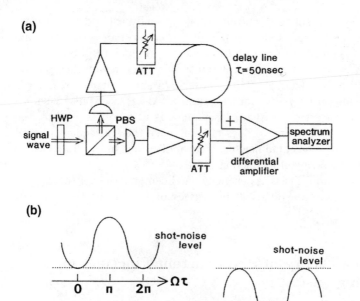

Fig. 1.5. (a) Balanced direct detectors with a delay line and attenuators (ATT). (b) Current noise spectra of amplitude-antisqueezed light and amplitude-squeezed light. HWP: half-wave plate; PBS: polarizing beam splitter

this case, the first and second terms of (1.32) are canceled each other out exactly, and the photocurrent power spectrum is equal to the full shot-noise value. This noise is generated by the beat between the coherent amplitude of the test field and the vacuum field incident from an open port of the beam splitter [163]. At a fluctuation frequency Ω_{out} satisfying $\Omega_{\text{out}}\tau = (2N+1)\pi$, the differential amplifier produces the sum signal, $i_1 + i_2$, of the two photodetector outputs. In this case, the beat between the test field and the vacuum field incident from an open port of the beam splitter is exactly canceled out, and the photocurrent power spectrum corresponds to the first and second terms of (1.32). In this way the test field intensity noise and the corresponding shot-noise values are simultaneously displayed on a spectrum analyzer with a period of $\Delta\Omega = 2\pi/\tau$.

When the test field is antisqueezed, i.e., super-Poissonian, the photocurrent power spectral density at Ω_{in} (shot noise) is smaller than that at Ω_{out} (test field intensity noise), as shown in Fig. 1.5b. On the other hand, when the test field is squeezed, i.e., sub-Poissonian, the photocurrent power spectral density at Ω_{in} is larger than that at Ω_{out}, as shown in Fig. 1.5c. Figure 1.6

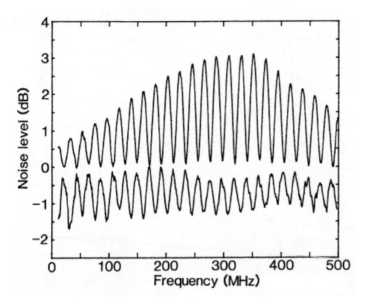

Fig. 1.6. Current noise spectra normalized by the shot-noise level from a constant-current-driven semiconductor laser for bias levels $I/I_{th} = 1.03$ (*upper curve*) and $I/I_{th} = 13.6$ (*lower curve*). The amplifier thermal noise is subtracted in the normalization process

shows an example of the photocurrent power spectra for the antisqueezed and squeezed fields, which are generated by the same (constant-current-driven) semiconductor laser, operated near to and far above the oscillation threshold, respectively [163].

2. Noise of p–n Junction Light Emitters

2.1 Introduction

It has been shown that the noise associated with injection of carriers into the active region of a light-emitting p–n junction ("pump noise") determines the intensity noise properties of the light output from these devices at low frequencies [262, 263]. Successful generation of intensity-squeezed light from a semiconductor laser [161] and sub-Poissonian light from a light-emitting diode (LED) [231] under constant-current driving conditions implies that the carrier injection into the active region can be regulated to well below the Poissonian limit.

Conventional noise theories [33, 239] of p–n junctions predict shot-noise-limited current fluctuations in the external circuit when the junction is biased with a constant-voltage source. Such a constant voltage driving condition, where the differential resistance, R_d, of the junction is much larger than the series resistance, R_s, of the external circuit, is difficult to achieve in a real experimental situation when the junction is driven with a large current. This is because the differential resistance of a p–n junction, $R_d \simeq k_B T/eI$, decreases as the DC current, I, increases. For example, when the p–n junction is operating at room temperature and the current through the junction is $I = 25$ mA, the differential resistance is $R_d = 1\,\Omega$. For p–n junctions operating in this regime, it is necessary to take into account the effect of external circuit impedance, R_s.

It has been demonstrated recently that electron flow in the presence of scattering centers features reduced shot noise, but still much larger than thermal noise [132, 158, 197], if the scattering mechanism is non-dissipative (or inelastic). Suppression of current shot noise (generation of "quiet electron flow") in a macroscopic resistor has been discussed in the context of mesoscopic physics [16, 156, 210]. This shot noise suppression by inelastic electron scattering in a source resistor provides the physical basis for the thermal-noise-limited current fluctuation [113, 181] used to describe the noise properties of a macroscopic resistor. Although the noise generated in the external resistor is far below the shot-noise level, this does not mean that the carrier injection into the active region of a p–n junction is regulated. In fact, it has been pointed out recently using a microscopic thermionic emission model that the carriers supplied by the external circuit are injected stochastically across the depletion layer before they reach the active region [103].

The noise properties of the current flowing in the external circuit connected to a p–n junction was analyzed by Yamamoto and Machida [263] and Yamamoto and Haus [266] using a macroscopic diffusion model. The authors discovered that the shot-noise-limited current fluctuation in a constant-voltage-driven p–n junction diode is not the pump noise for the p–n junction light emitter, but arises as a consequence of the random diffusion and generation–recombination processes inside the diode. Therefore, this does not explain the reason why the carrier injection across the depletion layer ("pumping process") is regulated by a constant-current driving condition. The physical mechanism that suppresses the pump noise was discovered when microscopic carrier motion across the depletion layer was considered. The charging energy at the junction plays a key role in establishing the correlation between successive carrier-injection events, thereby regulating this stochasticity [103].

In this chapter, a rather general noise theory of p–n junction light-emitting devices in the macroscopic regime is presented. The charging energy of carriers across the depletion layer is taken into account by considering the Poisson equation at the junction. The resulting carrier dynamics is analyzed by a set of Langevin equations. The noise and correlation properties of quantities like the generated photon field, external circuit current, junction voltage and carrier number in the active region can be calculated using this formalism [205]. This model provides a complete understanding of the noise properties of p–n junction light-emitting devices in the macroscopic limit, including the sub-shot-noise intensity fluctuations in a squeezed state and sub-Poissonian light generated by a semiconductor laser and an LED.

2.2 Junction Voltage Dynamics: the Poisson Equation

In order to treat the noise properties of light generated by a p–n junction, the active medium where the photons are generated, its inverted population, and the corresponding pumping mechanism have to be defined. Here, it is assumed that the junction current is mainly carried by the injection of electrons into the p-type layer and subsequent radiative recombination, so that the optically active region is formed in the p-type layer. Under this assumption, it is natural to define the active medium to be the region in the p-type layer where injected electrons recombine, and the inverted population to be the total number of excess minority carriers (electrons) compared to their equilibrium value in this region. The pumping process will be the injection of electrons into this region across the depletion layer from the n-type layer, which serves as the electron reservoir. It is further assumed that there is no carrier recombination within the depletion layer. The situation is illustrated in Fig. 2.1, where the two cases of a homojunction and a heterojunction are shown.

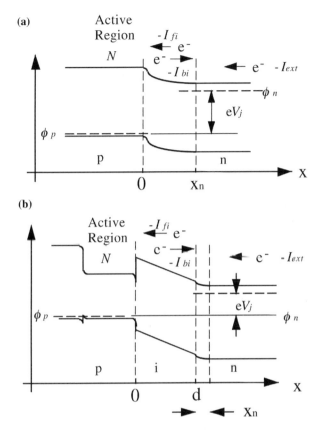

Fig. 2.1 a, b. Schematic of the junction and the parameters considered. *Arrows* indicate the direction of electron flow. (a) p–n homojunction; (b) p–i–n heterojunction

To describe the pumping process, one needs to analyze the carrier dynamics at the junction across the depletion layer. Throughout the analysis, it is assumed that (a) the carrier thermalization rate by phonon scattering is larger than any other rate, so that the electrons in the n-type layer are always in thermal equilibrium (b) the recombination lifetime (τ_{sp}) is short, so that the major current-flow mechanism is the radiative recombination of minority carriers. Assumption (a) guarantees that the carrier motion can be described by the semi-classical picture employed here. Assumption (b) holds in materials like GaAs with a direct bandgap.

The Poisson equation is given by

$$\nabla^2 V = -\rho/\epsilon \ , \tag{2.1}$$

where ρ is the space charge density and ϵ is the dielectric constant of the material. Integration of this equation gives

$$\frac{dV_j}{dt} = \frac{1}{C_{dep}} I_{ext}(t) \quad , \tag{2.2}$$

which describes the dynamics of the junction voltage $V_j(t)$, which is defined to be the difference between the quasi-Fermi levels of the n-type layer (ϕ_n) and the p-type layer (ϕ_p) at the junction.

First, consider an abrupt p–n homojunction (Fig. 2.1a) where the depletion layer is formed within the uniformly doped layer. If V_1 is the junction potential supported by the n-type layer and V_2 is that supported by the p-type layer, the total potential V_{tot} supported by the junction is given by $V_{tot} = V_1 + V_2$. In the limit of an one-sided junction where the doping level of the p-type layer is much higher than that of the n-type layer, $V_2 \ll V_1$ [223]. This assumption is valid for most GaAs/AlGaAs-based lasers and LEDs, where p-type doping is usually much higher than n-type doping. Integrating (2.1) gives the junction voltage V_j as

$$V_j = V_{bi} - V_{tot} = \frac{eN_D}{2\epsilon}(x_{n0}^2 - x_n^2) \quad . \tag{2.3}$$

Here, the space charge density ρ is given by the doping density eN_D. V_{bi} is the built-in potential, x_{n0} and x_n are the width of the depletion layer at zero bias and at finite bias V_j, respectively.

The space charge separated by the junction, Q, can be defined in terms of the width of the depletion layer and is given by $Q \equiv eN_D A_0 x_n$, where A_0 is the cross-sectional area of the junction. The depletion layer capacitance is defined by [223]

$$C_{dep} \equiv \left| \frac{dQ}{dV_j} \right| = \frac{\epsilon A_0}{x_n} \quad . \tag{2.4}$$

The rate of change of Q is equal to the circuit current $I_{ext}(t)$,

$$I_{ext}(t) \equiv -\frac{dQ}{dt} = -eN_D A_0 \frac{dx_n}{dt} \quad . \tag{2.5}$$

Using (2.3) and (2.5), one gets (2.2).

The case of a p–i–n heterojunction (Fig. 2.1b) was studied by Imamoḡlu et al. [101, 102] in a similar way. Equation (2.3) is replaced by

$$V_j = \frac{eN_D}{2\epsilon}[x_{n0}^2 - x_n^2 + 2d(x_{n0} - x_n)] \quad , \tag{2.6}$$

where d is the intrinsic layer width and x_n is the depletion layer width in the n-type layer. Equation (2.2) still holds for this case, with a slightly different definition of the depletion layer capacitance, $C_{dep} = \epsilon A_0/(d + x_n)$.

It is generally possible to write down (2.2) for any p–n junction structures with an appropriate definition of the depletion layer capacitance. There are three mechanisms that contribute to the change in the depletion layer width, x_n. The first is the current flowing in the external circuit, which pushes the electron cloud forward and thus decreases the width of the depletion layer until the steady-state junction voltage is established. The second is the forward injection of electrons into the active layer across the depletion layer. This forward-injection mechanism increases the space charge at both sides of the depletion layer and increases the depletion layer width x_n. The third is the backward injection of electrons from the active region back to the n-type layer across the depletion layer, which decreases the depletion layer width x_n. One can show that the effect of the fluctuation in the depletion layer capacitance on the current is small compared to that due to junction voltage fluctuation; the depletion layer capacitance, C_{dep}, is considered to be constant under cw operation.

2.3 Semiclassical Langevin Equation for Junction Voltage Dynamics

The forward-injection mechanism is modeled as diffusion of electrons across the depletion layer from the n-type layer to the p-type layer. The average forward-injection diffusion current is given by $I_{fi} = \frac{1}{2}en_p v A_0$, where n_p is the electron density at $x = 0$ and v is thermal velocity of the electron in the $-x$ direction, given by l_f/τ_f [178]. Here, l_f is the electron mean-free-path and τ_f is the mean-free-time. Using $n_p = n_{p0}\exp[eV_j/k_BT]$ and the Einstein relation $D_n = l_f^2/2\tau_f$, where n_{p0} is the equilibrium electron concentration in the p-type layer and D_n is the electron diffusion constant, one gets [34, 33]

$$I_{fi}(t) = \frac{en_{p0}D_nA_0}{l_f}\exp\left(\frac{eV_j(t)}{k_BT}\right) \quad . \tag{2.7}$$

The minority carrier distribution for $x < 0$, in the presence of radiative recombination with carrier lifetime of τ_{sp}, is given by $n_p(x) = n_{p0} + (n_p - n_{p0})\exp(x/L_n)$ with $L_n^2 = D_n\tau_{sp}$ [224]. The total number of excess electrons in the active region is given by integrating $n_p(x) - n_{p0}$ from $x = -\infty$ to $x = 0$

$$N = A_0L_n(n_p - n_{p0}) \quad . \tag{2.8}$$

Since the thermal motion is random, there are carriers that are injected back from the active region into the n-type layer. On average, this backward-injection current is proportional to the carrier flux in the $+x$ direction from $x = -l_f$,

$$I_{bi}(t) = \frac{eD_nA_0}{l_f}\left(n_{p0} + \frac{N}{A_0L_n}\exp(-l_f/L_n)\right) \quad . \tag{2.9}$$

The net diffusion current I is given by the difference between the forward- and backward-injection currents. Expanding $\exp(-l_f/L_n) \simeq 1 - l_f/L_n$ yields

$$I = \frac{eD_nA_0}{L_n}n_{p0}[\exp(eV_j/k_BT) - 1] = I_0[\exp(eV_j/k_BT) - 1] \quad , \tag{2.10}$$

where $I_0 = eD_\mathrm{n}A_0 n_\mathrm{p0}/L_\mathrm{n}$ is the reverse saturation current. This is the electrical diffusion current first obtained by Shockley [212], which is seen as the external circuit current. However, one needs to consider the full dynamics of I_fi and I_bi, since they constitute the microscopic pumping process.

Since the forward- and backward-injection currents change the depletion layer width, they will affect the junction voltage according to (2.3) or (2.6). After taking the effects of these currents into account, (2.2) reduces to

$$\frac{dV_\mathrm{j}}{dt} = \frac{1}{C_\mathrm{dep}} I_\mathrm{ext}(t) - \frac{1}{C_\mathrm{dep}} I_\mathrm{fi}(t) + \frac{1}{C_\mathrm{dep}} I_\mathrm{bi}(t) \quad . \tag{2.11}$$

Let us consider the diffusion (or thermionic emission) of an electron across the depletion layer as a discrete and instantaneous process. Imagine a forward-injection event occurring at time $t = t_0$. The junction voltage drop due to the forward-injection process is $\Delta V_\mathrm{j} = e/C_\mathrm{dep}$. This will, on average, decrease the forward-injection current by a factor

$$\begin{aligned}
\frac{I_\mathrm{fi}(t = t_0+)}{I_\mathrm{fi}(t = t_0-)} &= \frac{I_0 \exp[eV_\mathrm{j}(t = t_0+)/k_\mathrm{B}T]}{I_0 \exp[eV_\mathrm{j}(t = t_0-)k_\mathrm{B}T]} \\
&= \exp\left(-\frac{e}{k_\mathrm{B}T}\,\Delta V_\mathrm{j}\right) \\
&\equiv \exp(-r) \quad ,
\end{aligned} \tag{2.12}$$

where $r \equiv \left(e^2/C_\mathrm{dep}\right)/k_\mathrm{B}T$, the ratio between the single-electron charging energy and thermal energy. It should be noted that this is a comparison of the single-electron charging energy e^2/C_dep with the characteristic energy scale to change the forward-injection current significantly, which for the case of a p–n junction happens to coincide with the thermal energy, $k_\mathrm{B}T$. Two cases are considered, when $r \gg 1$ and when $r \ll 1$.

2.3.1 Mesoscopic Case ($r \gg 1$)

In the case of extremely low temperature ($T \sim 50$ mK) and small capacitance ($C_\mathrm{dep} \sim 1$ fF), the condition $r \gg 1$ is satisfied. In this limit, a single carrier-injection event will decrease the forward-injection current virtually to zero, and any subsequent injection event is drastically suppressed. The injection is in general not stochastic, and (2.11) cannot be linearized. An analytical solution is not available, but the situation can be simulated by employing the Monte Carlo method [101, 102, 103]. The theoretical treatment and experimental realization of a p–n junction in this mesoscopic limit are discussed in Chapters 10 and 11, respectively.

2.3.2 Macroscopic Case ($r \ll 1$)

For a normal laser diode or LED with a large capacitance, C_dep, operating at reasonably high temperature (≥ 4 K), the factor r is much smaller than unity.

Under this condition, a single carrier-injection event does not change the average forward-injection current appreciably. Therefore, one has a stochastic injection process with the average rate determined by the junction voltage. In this limit, one can split the forward-injection current into two parts: an average current which varies only as a function of the time-dependent junction voltage and a stochastic noise current due to individual random injection events which have zero average. The forward-injection current term can be written as

$$I_{\mathrm{fi}}(t) = \frac{eD_{\mathrm{n}}A_0}{l_{\mathrm{f}}} \exp\left(\frac{eV_{\mathrm{j}}(t)}{k_{\mathrm{B}}T} \right) + e\,F_{\mathrm{fi}} \quad, \tag{2.13}$$

where $V_{\mathrm{j}}(t)$ denotes a time-dependent junction voltage and F_{fi} is the Langevin noise term. Since F_{fi} arises from collisions with a reservoir of phonons and other electrons which have large degrees of freedom, the correlation time is infinitesimally short. Therefore, one obtains the Markovian correlation function

$$\langle F_{\mathrm{fi}}(t)F_{\mathrm{fi}}(t') \rangle = \frac{2\langle I_{\mathrm{fi}}(t) \rangle}{e} \delta(t - t') \quad, \tag{2.14}$$

where $\langle I_{\mathrm{fi}}(t) \rangle$ denotes the slowly varying average forward-injection current. One can make similar arguments for the backward-injection current, and thus obtain

$$I_{\mathrm{bi}}(t) = \frac{eD_{\mathrm{n}}A_0}{l_{\mathrm{f}}} \left(n_{\mathrm{p}0} + \frac{N}{A_0 L_{\mathrm{n}}} \exp(-l_{\mathrm{f}}/L_{\mathrm{n}}) \right) + e\,F_{\mathrm{bi}} \quad, \tag{2.15}$$

with

$$\langle F_{\mathrm{bi}}(t)F_{\mathrm{bi}}(t') \rangle = \frac{2\langle I_{\mathrm{bi}}(t) \rangle}{e} \delta(t - t') \quad. \tag{2.16}$$

To describe the effect of different driving conditions, consider the case where the p–n junction is connected to a constant voltage source, with a series resistor, R_{s}, that carries thermal voltage noise, V_{s} (Fig. 2.2a). The forward- and backward-injection events are described in the equivalent circuit model (Fig. 2.2b) as independent current sources (I_{fi} and I_{bi}) charging or discharging the depletion layer capacitor C_{dep}. Defining $F_{\mathrm{rs}} \equiv V_{\mathrm{s}}/eR_{\mathrm{s}}$, the external current is given by

$$I_{\mathrm{ext}}(t) = \frac{V - V_{\mathrm{j}}}{R_{\mathrm{s}}} + e\,F_{\mathrm{rs}} \quad, \tag{2.17}$$

with

(a)

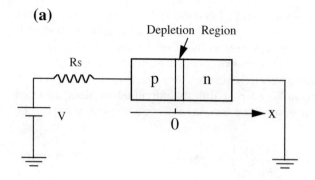

(b)

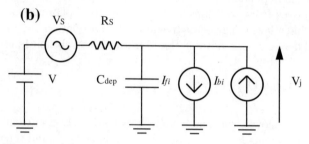

Fig. 2.2. (a) A p–n junction diode connected in series with a resistor R_s that carries voltage noise of V_s, altogether biased by a constant-voltage source, V. (b) Equivalent circuit model

$$\langle F_{rs}(t) F_{rs}(t') \rangle = \frac{4 k_B T}{e^2 R_s} \delta(t - t') \quad , \tag{2.18}$$

and (2.11) reduces to

$$\frac{dV_j}{dt} = \frac{V - V_j}{R_s C_{dep}} - \frac{I_{fi}(V_j)}{C_{dep}} + \frac{I_{bi}(N)}{C_{dep}} + \frac{e}{C_{dep}}(-F_{fi} + F_{bi} + F_{rs}) . \tag{2.19}$$

This equation describes the suppression of the pump noise in conventional semiconductor p–n junction light emitters in the macroscopic regime.

2.4 Noise Analysis of an LED

In this section, the semi-classical Langevin equations that describe the noise properties of an LED are considered. Use of semi-classical equations is justified since an LED does not have an optical feedback mechanism to generate phase coherent light. Figure 2.3 shows the model of the LED under consideration. The carriers are injected into the active layer, and the junction voltage fluctuation is described by (2.19). The total number of minority carriers (electrons) N in the active p-type layer increases by forward-injection current and

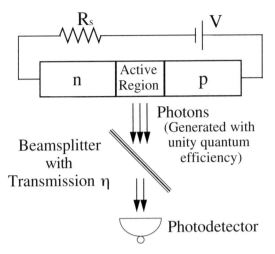

Fig. 2.3. Model of an LED. The carriers are pumped into the active region, and the photons are generated by spontaneous emission with unity quantum efficiency. Any internal loss or collection imperfection is modeled by a beam splitter with transmission probability η, with $0 \leq \eta \leq 1$

spontaneous absorption, and it decreases due to backward-injection current and radiative recombination. The spontaneous absorption and recombination are described by a Poisson point process with a fixed lifetime of τ_{sp}. Optical losses (either internal or external) are modeled as a beam splitter with transmission probability η ($0 \leq \eta \leq 1$). The equations that describe the system are

$$\frac{C_{\mathrm{dep}}}{e}\frac{\mathrm{d}V_{\mathrm{j}}}{\mathrm{d}t} = \frac{V - V_{\mathrm{j}}}{eR_{\mathrm{s}}} - \frac{I_{\mathrm{fi}}(V_{\mathrm{j}})}{e} + \frac{I_{\mathrm{bi}}(N)}{e} - F_{\mathrm{fi}} + F_{\mathrm{bi}} + F_{\mathrm{rs}} , \quad (2.20)$$

$$\frac{\mathrm{d}N}{\mathrm{d}t} = -\frac{N + N_{\mathrm{p0}}}{\tau_{\mathrm{sp}}} + \frac{N_{\mathrm{p0}}}{\tau_{\mathrm{sp}}} + \frac{I_{\mathrm{fi}}(V_{\mathrm{j}})}{e} - \frac{I_{\mathrm{bi}}(N)}{e} - F_{\mathrm{sp}} + F_{\mathrm{fi}} - F_{\mathrm{bi}} , \quad (2.21)$$

$$\Phi = \eta\left[\frac{N + N_{\mathrm{p0}}}{\tau_{\mathrm{sp}}} - \frac{N_{\mathrm{p0}}}{\tau_{\mathrm{sp}}} + F_{\mathrm{sp}}\right] + F_{\mathrm{v}} . \quad (2.22)$$

$N_{\mathrm{p0}}/\tau_{\mathrm{sp}} = n_{\mathrm{p0}}A_0L_{\mathrm{n}}/\tau_{\mathrm{sp}} = I_0/e$ gives the rate of spontaneous absorption or recombination due to background electron concentration in the active region. F_{sp} is the noise term corresponding to spontaneous absorption and radiative recombination. Since these processes arise from coupling with vacuum fluctuations in an infinite number of modes, the correlation of F_{sp} becomes

$$\langle F_{\mathrm{sp}}(t)F_{\mathrm{sp}}(t')\rangle = 2\frac{N + 2N_{\mathrm{p0}}}{\tau_{\mathrm{sp}}}\delta(t - t') . \quad (2.23)$$

The noise source corresponding to forward injection, F_{fi}, appears in the equations for both the carrier number (2.21) and the junction voltage (2.20) and the two terms are negatively correlated, so one is replaced with the negative of the other. The same is true for F_{bi}. Φ is the photon flux measured at the photodetector, after the photons pass through a beam splitter of transmission probability η. The beam splitter introduces partition noise, F_{v}, into the detector, which has a Markovian correlation of

$$\langle\, F_{\mathrm{v}}(t)F_{\mathrm{v}}(t')\,\rangle \;=\; 2\eta(1-\eta)\,\frac{N+2N_{\mathrm{p0}}}{\tau_{\mathrm{sp}}}\,\delta(t-t') \quad , \tag{2.24}$$

since the noise comes from the vacuum fluctuations coupling in from the open port of the beam splitter.

2.4.1 Steady-State Conditions

In the steady state, (2.21) and (2.20) yield

$$\frac{N_0}{\tau_{\mathrm{sp}}} \;=\; \frac{I_{\mathrm{fi}}(V_0)}{e} - \frac{I_{\mathrm{bi}}(N_0)}{e} \;=\; \frac{V-V_0}{eR_{\mathrm{s}}} \quad , \tag{2.25}$$

where the subscript "0" denotes the steady-state values. If the operation current $I \equiv (V-V_0)/(eR_{\mathrm{s}})$ is given, (2.25) can be used to determine the following quantities uniquely:

- N_0 is determined by $N_0 = \frac{V-V_0}{eR_{\mathrm{s}}}\tau_{\mathrm{sp}}$,
- Φ_0 is determined by $\Phi_0 = \eta\frac{N_0}{\tau_{\mathrm{sp}}}$,
- N_0 determines $I_{\mathrm{bi}}(N_0)$, and
- $I_{\mathrm{fi}}(V_0) - I_{\mathrm{bi}}(N_0) = \frac{V-V_0}{eR_{\mathrm{s}}}$ determines $I_{\mathrm{fi}}(V_0)$ and thus V_0.

2.4.2 Linearization

Once the steady-state conditions are determined, one can linearize the equations around these steady-state values (Appendix A.1), using

$$N = N_0 + \Delta N, \tag{2.26}$$
$$V_{\mathrm{j}} = V_0 + \Delta V, \tag{2.27}$$
$$\Phi = \Phi_0 + \Delta\Phi. \tag{2.28}$$

One needs to evaluate two time constants, τ_{fi} and τ_{bi}, defined by (A.4) and (A.5) respectively. The diffusion model is employed for forward- and backward-injection currents, as discussed in Sect. 2.3. From (2.7) and (A.4) one gets

$$\frac{1}{\tau_{\mathrm{fi}}} - \frac{1}{C_{\mathrm{dep}}}\frac{\mathrm{d}}{\mathrm{d}V_{\mathrm{j}}}I_{\mathrm{fi}}(V_{\mathrm{j}})\Big|_{V_{\mathrm{j}}=V_0} \;=\; \frac{eI_{\mathrm{h}}(V_0)}{k_{\mathrm{B}}TC_{\mathrm{dep}}} \quad . \tag{2.29}$$

For the backward-injection current, (2.9) and (A.5) gives

$$\frac{1}{\tau_{bi}} = \frac{1}{e}\frac{d}{dN}I_{bi}(N)\Big|_{N=N_0} = \frac{I_{bi}(N_0)}{e[N_0 + n_{p0}A_0L_n\exp(l_f/L_n)]} \quad . \quad (2.30)$$

Since the electron mean-free-path l_f is much smaller than the electron diffusion length L_n, at reasonably high bias, $eV_j/k_BT \gg 1$, the relation

$$I_{fi}(V_0) \simeq I_{bi}(N_0) \gg I \quad (2.31)$$

is obtained, with $I = I_{fi}(V_0) - I_{bi}(N_0)$, where $I_{fi}(V_0)$ and $I_{bi}(N_0)$ are the average forward- and backward-injection currents, respectively. The time constants τ_{fi} and τ_{bi} represent the characteristic time scales of thermal fluctuations. Since these are the shortest time scales in the problem, one finds the condition

$$\tau_{fi}, \tau_{bi} \ll \tau_{sp}, \tau_{te}, \tau_{RC} \quad . \quad (2.32)$$

We call this the diffusion limit.

For mesoscopic p–n junctions, considered by Imamoğlu et al. [101, 102, 103], it is assumed that the carrier-injection mechanism is thermionic emission across the potential barrier. Those authors assumed that the carrier recombines immediately after it is injected and that there is little accumulation of carriers in the active layer ($\tau_{sp} \ll e/I$). Here, the forward-injection carries the net junction current, and the backward-injection current is negligible. The model described here can characterize the junction in this limit by considering the conditions

$$I_{fi}(V_0) \simeq I \gg I_{bi}(N_0) \quad , \quad (2.33)$$

$$\tau_{fi}, \tau_{sp}, \tau_{RC} \ll \tau_{bi} \quad . \quad (2.34)$$

We call this the thermionic emission limit. Although this case can be described by the model, the noise analysis in the diffusion limit is of most interest here.

2.4.3 Photon-Number Noise

Linearizing the equations and taking the Fourier transform allows us to calculate the intensity noise power spectral density of the photons detected at the photodetector (Appendix A.2).

The "thermionic emission time", τ_{te}, is defined as [101, 102, 103, 122]

$$\tau_{te} \equiv R_d C_{dep} = \frac{k_B T C_{dep}}{e(I + I_0)} \simeq \tau_{fi}\tau_{sp}/\tau_{bi} \quad . \quad (2.35)$$

The last equality follows from $I = I_{fi}(V_0) - I_{bi}(N_0)$ and (2.25), (2.29), (2.30) and (2.32). τ_{te} is the time scale over which V_j fluctuates by k_BT/e. Using this definition and condition (2.32), $\chi(\Omega)$ defined in (A.17) becomes

$$\chi(\Omega) \simeq \left[\left(1 + \frac{\tau_{\text{te}}}{\tau_{\text{RC}}} \right)^2 + \Omega^2 (\tau_{\text{sp}} + \tau_{\text{te}})^2 \right]^{-1} , \qquad (2.36)$$

and (A.16) reduces to

$$S_{\Delta\Phi} = \eta \frac{2I}{e} (1 - \eta\chi(\Omega)) + \eta \frac{4k_{\text{B}}T}{e^2 R_{\text{d0}}} \left[1 - \eta\chi(\Omega) \left(1 + \frac{\tau_{\text{te}}}{\tau_{\text{RC}}} \right) \right] , \quad (2.37)$$

where $R_{\text{d0}} = (dV_{\text{j}}/dI)|_{V_{\text{j}}=0}$ is the differential resistance of the junction at zero bias.

When the junction is driven by a constant-current source, the condition $\tau_{\text{RC}} \gg \tau_{\text{sp}}, \tau_{\text{te}}$ is satisfied. In this case, (2.37) further reduces to

$$S_{\Delta\Phi} \rightarrow \eta \left(\frac{2I}{e} + \frac{4k_{\text{B}}T}{e^2 R_{\text{d0}}} \right) \left(1 - \frac{\eta}{1 + \Omega^2 (\tau_{\text{sp}} + \tau_{\text{te}})^2} \right) . \qquad (2.38)$$

The photon-number noise is reduced to below the shot-noise value at frequencies low compared to $1/(\tau_{\text{sp}} + \tau_{\text{te}})$. At very low frequencies, the photon noise is $1 - \eta$. At high frequencies, the photon noise approaches the shot-noise value. This is in agreement with experimental observation [77, 122, 231].

When the junction is driven by a constant voltage source, $\tau_{\text{RC}} \ll \tau_{\text{sp}}, \tau_{\text{te}}$ and

$$S_{\Delta\Phi} \rightarrow \eta \left(\frac{2I}{e} + \frac{4k_{\text{B}}T}{e^2 R_{\text{d0}}} \right) . \qquad (2.39)$$

In this case, the photon is shot noise limited at high bias ($I \gg I_0$) and reduces to the thermal noise of the junction at zero bias.

Figure 2.4a shows the intensity noise power spectral density (2.37) of an LED under high bias ($I \gg I_0$). The junction parameters, like the depletion layer capacitance C_{dep} and temperature T, are fixed and the current is adjusted so that $\tau_{\text{te}} = \tau_{\text{sp}}$. One can see that as the source resistance R_{s} (and thus the time constant of the circuit τ_{RC}) is increased, the noise at low frequencies is reduced to below the Poisson limit. The effect of finite quantum efficiency ($\eta = 0.5$) is shown in Fig. 2.4b. The ultimate squeezing level is determined by the imperfect quantum efficiency.

One can define the squeezing bandwidth to be the frequency at which the degree of squeezing is reduced by a factor of 2 compared to the squeezing at zero frequency. This can be explicitly calculated from (2.38) to give

$$f_{\text{3dB}} = \frac{\Omega_{\text{3dB}}}{2\pi} = \frac{1}{2\pi(\tau_{\text{sp}} + \tau_{\text{te}})} = \frac{1}{2\pi \left(\tau_{\text{sp}} + \frac{k_{\text{B}}TC_{\text{dep}}}{e(I+I_0)} \right)} . \qquad (2.40)$$

In Fig. 2.5, the noise power spectral density under different driving currents is plotted, and the dependence of squeezing bandwidth on current is shown. It should be noted that the squeezing bandwidth is affected neither by the value of source resistance (Fig. 2.4) nor by the quantum efficiency

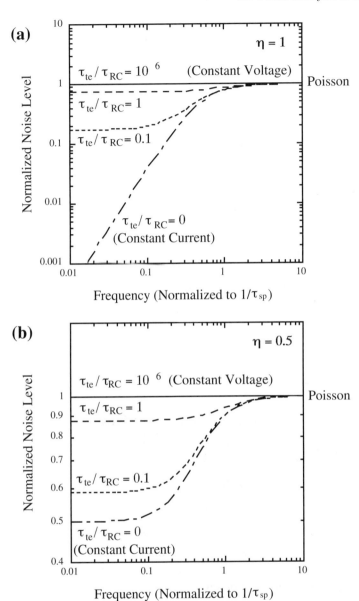

Fig. 2.4. LED noise power spectral density calculated for a junction under different bias conditions, at high bias ($I \gg I_0$). Operation current is chosen so that $\tau_{te} = \tau_{sp}$. The bias condition is treated by changing the series resistor R_s, and thus the time constant τ_{RC}. (a) Unity quantum efficiency ($\eta = 1$); (b) $\eta = 0.5$

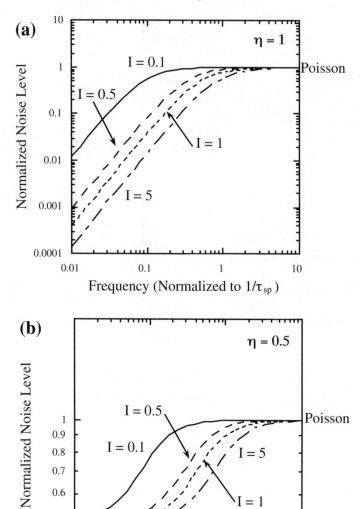

Fig. 2.5 a, b. LED noise power spectral density calculated for a junction driven by a constant current source, at high bias ($I \gg I_0$). Frequency is normalized to $1/\tau_{\mathrm{sp}}$ and current is normalized to $I_{\mathrm{N}} = e/(r\tau_{\mathrm{sp}})$, so that $I = I_{\mathrm{N}}$ corresponds to $\tau_{\mathrm{te}} = \tau_{\mathrm{sp}}$. (a) Unity quantum efficiency ($\eta = 1$); (b) $\eta = 0.5$

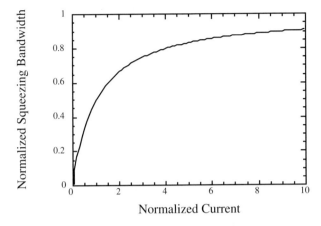

Fig. 2.6. Squeezing bandwidth of an LED plotted as a function of driving current. The bandwidth is normalized to $1/(2\pi\tau_{\rm sp})$, and driving current is normalized to $I_{\rm N} = e/(r\tau_{\rm sp})$, so that $I = I_{\rm N}$ corresponds to $\tau_{\rm te} = \tau_{\rm sp}$

(Figs. 2.4b and 2.5b). Equation (2.40) describes the experimental result well [1, 122, 277]. Figure 2.6 shows the squeezing bandwidth as a function of driving current. The current is normalized so that $I = 1$ corresponds to $\tau_{\rm te} = \tau_{\rm sp}$.

2.4.4 Noise in the External Circuit Current

The linearized Langevin equations allow one to calculate the power spectrum of noise in the external circuit current (Appendix A.3). From (A.19) one can show that the external current noise approaches the thermal noise $4k_{\rm B}T/R_{\rm s}$ in the external resistor with a constant-current source ($\tau_{\rm RC} \gg \tau_{\rm sp}, \tau_{\rm te}$). With a constant-voltage source ($\tau_{\rm RC} \ll \tau_{\rm sp}, \tau_{\rm te}$), the noise approaches $2eI + 4k_{\rm B}T/R_{\rm d0}$. At high bias ($I \gg I_0$) the external current carries the full shot noise, and at zero bias it carries the thermal noise of the diode.

Figure 2.7 shows the schematic dynamics of a stochastic photon emission event, the junction voltage fluctuation, and the relaxation current in the external circuit in the diffusion limit. Sect. 2.4.5 shows that the correlation between the junction voltage and carrier number in the active region is perfect for macroscopic p–n junctions in the diffusion limit. A photon emission event accompanies a reduction in the carrier number, which creates a junction voltage drop of $e/C_{\rm dep}$. This fluctuation in the voltage is recovered by a relaxation current flow in the external circuit with a time scale of $\tau_{\rm RC}$. When the junction is driven with a constant-voltage source (Fig. 2.7a), the junction voltage (and thus the carrier number) recovers very quickly, and the next emission event is independent of the previous event. The photon-emission event is a Poisson point process, and the relaxation current flows accordingly.

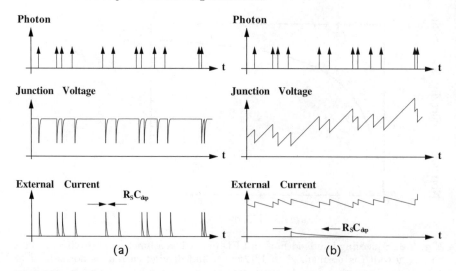

Fig. 2.7 a, b. Schematic showing the photon emission, junction voltage dynamics, and external current flow of a p–n junction. (a) Constant-voltage-driven case; (b) constant-current-driven case

The external current therefore features the full shot noise. In the constant-current-driven case (Fig. 2.7b), the relaxation current flows very slowly; thus, the next photon-emission event occurs before the external circuit completely recovers the junction voltage. The external circuit current due to the second emission event is superimposed on the first one, and the resulting fluctuation is less than the shot noise.

It is rather interesting to note the origin of noise in the external circuit current. The carriers jump back and forth across the depletion layer and establish the correlation between the junction voltage and the carrier number. However, these events do not contribute to the external current noise, because the effective resistance across the junction depletion layer ($k_B T/eI_{\mathrm{fi}}$, $k_B T/eI_{\mathrm{bi}}$) is so small that the junction voltage drop induced by a forward- (backward-) injection event will be relaxed mostly by a direct backward- (forward-) injection event rather than through the external circuit. This means that the forward- and backward- injection events will not be seen from the external circuit and all the noise will come from the recombination event in the active region.

2.4.5 Correlation Between Carrier Number and Junction Voltage

The normalized correlation between the junction voltage fluctuation and the carrier number fluctuation is defined as

$$C_{n,v}(\Omega) \equiv \frac{\langle \frac{C_{\text{dep}}}{e} \Delta \tilde{N}^*(\Omega) \Delta \tilde{V}_j(\Omega) \rangle}{\langle \Delta \tilde{N}^*(\Omega) \Delta \tilde{N}(\Omega) \rangle^{\frac{1}{2}} \langle \frac{C_{\text{dep}}}{e} \Delta \tilde{V}_j^*(\Omega) \frac{C_{\text{dep}}}{e} \Delta \tilde{V}_j(\Omega) \rangle^{\frac{1}{2}}} . \quad (2.41)$$

where $*$ denotes the complex conjugate. One can calculate the correlation coefficient $|C_{v,n}|$ from the linearized Langevin equations of Appendix A.1 (Appendix A.4).

One can see from (A.20 – A.22) that $|C_{v,n}|^2$ approaches 1 and there is a perfect correlation between the junction voltage and the carrier number in the active region no matter what the driving condition is (i.e., for all values of τ_{RC}). The physical reason behind this is the fast (forward- and backward-) injection events which quickly restore a unique relation between carrier number fluctuation and junction voltage fluctuation with the characteristic relaxation time $\sim \tau_{fi}$, τ_{bi}. Since this relaxation is so fast, fluctuation of one of the two variables will immediately be followed by fluctuation of the other.

2.4.6 Correlation Between Photon Flux and Junction Voltage

The normalized correlation between the photon flux and junction voltage fluctuation is defined as

$$C_{\Phi,v}(\Omega) \equiv \frac{\langle \Delta \tilde{\Phi}^*(\Omega) \frac{C_{\text{dep}}}{e} \Delta \tilde{V}_j(\Omega) \rangle}{\langle \Delta \tilde{\Phi}^*(\Omega) \Delta \tilde{\Phi}(\Omega) \rangle^{\frac{1}{2}} \langle \frac{C_{\text{dep}}}{e} \Delta \tilde{V}_j^*(\Omega) \frac{C_{\text{dep}}}{e} \Delta \tilde{V}_j(\Omega) \rangle^{\frac{1}{2}}} , \quad (2.42)$$

which one can calculate similarly (Appendix A.5).

With (A.22-A.24), $|C_{\Phi,v}(\Omega)|^2$ reduces to

$$|C_{\Phi,v}(\Omega)|^2 \simeq$$
$$\frac{(\tau_{te}/\tau_{RC})^2 + \Omega^2(\tau_{sp} + \tau_{te})^2}{(1 + 2\tau_{te}/\tau_{RC})\left\{\frac{1}{\eta}[(1 + \tau_{te}/\tau_{RC})^2 + \Omega^2(\tau_{sp} + \tau_{te})^2] - 1\right\}} . \quad (2.43)$$

When the junction is driven with a constant-voltage source, this correlation reduces to 0, because the junction voltage fluctuation is merely determined by the noise in the external resistor.

When the junction is driven with a constant-current source, the correlation reduces to

$$|C_{\Phi,v}(\Omega)|^2 \rightarrow \frac{\Omega^2(\tau_{sp} + \tau_{te})^2}{\frac{1}{\eta}[1 + \Omega^2(\tau_{sp} + \tau_{te})^2] - 1} . \quad (2.44)$$

Whenever the photon is emitted, the junction voltage drops by e/C_{dep}, but it takes a long time for the external circuit to recover the junction voltage. If the photon is lost due to finite quantum efficiency, the junction voltage will decrease without a photon being detected, resulting in a decrease in correlation. This results in a correlation of η at high frequencies [compared

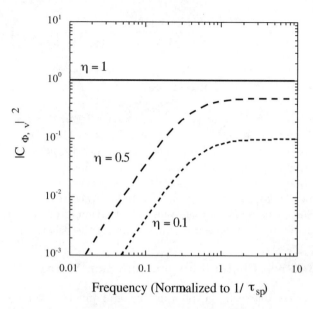

Fig. 2.8 a, b. Normalized correlation $|C_{\Phi,\mathrm{v}}|^2$ for several values of η when the LED is driven with a constant-current source

to $1/(\tau_{\mathrm{sp}} + \tau_{\mathrm{te}})]$. As the observation time gets longer (at lower frequencies), a second photon can be emitted after $\sim \tau_{\mathrm{sp}}$ or the junction voltage will fluctuate by approximately $k_{\mathrm{B}}T/e$ over time $\sim \tau_{\mathrm{te}}$. This results in the loss of correlation at lower frequencies. This is illustrated in Fig. 2.8, where the correlation is shown for several values of η.

2.5 Summary

In this chapter, a general model was presented to describe the noise properties of a macroscopic p–n junction light-emitting device. Langevin equations that describe the junction voltage, carrier number, and photon flux can be linearized in the macroscopic regime, since the effect of a single carrier does not dramatically modify the junction parameters. The model enables one to calculate the noise power spectral densities and correlations of various junction parameters such as output photon flux, external current fluctuation, junction voltage fluctuation and carrier number fluctuation. This successfully describes the generation of a sub-Poissonian photon flux under constant-current driving condition and recovers the predictions of conventional noise theories in the appropriate limits. It also predicts the bandwidth of squeezing quantitatively. In Chap. 3, an experiment is described that measures the squeezing bandwidth of an LED and confirms the predictions of this model.

3. Sub-Poissonian Light Generation in Light-Emitting Diodes

3.1 Introduction

In Chap. 2, a noise model was presented that described the noise properties of a semiconductor p–n junction light emitter in the macroscopic limit. One of the predictions made by the theory is for the squeezing bandwidth over which the intensity noise of the light output is suppressed to below the shot-noise level. In this chapter, the time scales involved in the pump-noise suppression mechanism, and how they determine the bandwidth of squeezing, are discussed. An experiment is presented where the squeezing bandwidth of the sub-Poissonian light generated by a LED is measured. The dependence of the squeezing bandwidth on the current, capacitance and temperature are consistent with the prediction made in the model presented in Chap. 2.

3.2 Physical Mechanism of Pump-Noise Suppression

The introduction of Chap. 2 discussed that the suppression of noise in the external circuit current does not guarantee regulated carrier-injection into an active layer across a depletion layer potential barrier. This is because the individual carrier injection is a random point process, the rate of which is determined purely by the junction voltage and the temperature of the junction. However, the (average) injection rate can be affected by the carrier injection events themselves. When a carrier is injected, the space charge in the depletion layer capacitance increases by e, and this decreases the junction voltage by e/C_{dep}. This decrease in the junction voltage decreases the carrier-injection rate, establishing a negative feedback mechanism to suppress the noise in the carrier-injection process. This mechanism is implemented mathematically in the model presented in Chap. 2 as the second and third terms in (2.11). If the junction (depletion layer) capacitance, C_{dep}, is large and/or the operation temperature, T, is high, the junction voltage drop, e/C_{dep}, due to a single carrier-injection event is much smaller than the thermal fluctuation voltage, $k_{\mathrm{B}}T/e$, so that the individual carrier-injection event does not influence the following event. We have a completely random point process (Poissonian carrier-injection process) with a constant rate in such a macroscopic p–n junction at a high temperature, even though the junction is driven by a "perfect constant-current source" [103]. This is in sharp contrast to a mesoscopic M–I–M tunnel junction [12, 66] and p–n junction [101, 102] at a low temperature, in which e/C_{dep} is much larger than $k_{\text{B}}T/e$ and the individual carrier-injection event is regulated by a fixed time separation e/I (single-electron Coulomb blockade effect).

Even in the macroscopic, high-temperature limit, the collective behavior of many electrons charging the depletion layer capacitance, C_{dep}, can amount to establishing regularity in the carrier-injection process. A single electron injection event reduces the junction voltage by e/C_{dep}, so the successive injection of $N = \left(\frac{k_{\text{B}}T}{e}\right) / \left(\frac{e}{C_{\text{dep}}}\right) = k_{\text{B}}TC_{\text{dep}}/e^2$ electrons reduces the junction voltage by the thermal voltage $k_{\text{B}}T/e$. Such a change in the junction voltage can result in significant modification of the carrier-injection rate. It is important to find out the time scale over which such a regulation mechanism works, since it will determine the effective bandwidth of this negative-feedback mechanism. Since carriers are provided by the external circuit at a rate of I/e, the time necessary for N carriers to be supplied can be calculated as

$$\tau_{\text{te}} = \frac{k_{\text{B}}TC_{\text{dep}}}{eI} = N\tau \quad , \tag{3.1}$$

where $\tau = \frac{e}{I}$ is the single-electron charging time. This time-scale is named "thermionic emission time", and is identical to (2.35) in the high-current limit ($I \gg I_0$). The junction current follows $\sim \exp(eV_{\text{j}}/k_{\text{B}}T)$, where V_{j} is the junction voltage. Therefore, τ_{te} is the time scale over which the junction current changes significantly. A carrier-injection event is completely stochastic at the microscopic level, but the junction voltage modulation induced by many carriers ($\sim N$) collectively regulates a global carrier-injection process over the time scale τ_{te}. If the measurement time, T_{meas}, is much longer than τ_{te}, the electron-injection process becomes sub-Poissonian. This is because the continuous charging or successive injection of N electrons modulates the junction voltage greater than $k_{\text{B}}T/e$, and, therefore, influences the subsequent events. If the measurement time, T_{meas}, is longer than τ_{te}, the variance of the injected electrons is given by the following fundamental limit [103]:

$$\langle \Delta n_{\text{e}}^2 \rangle = \frac{k_{\text{B}}TC_{\text{dep}}}{e^2} = N \quad . \tag{3.2}$$

This variance is independent of both the measurement time, T_{meas}, and the average electron number, $\langle n_{\text{e}} \rangle = \frac{I}{e}T_{\text{meas}}$. This independence is at the heart of squeezed light generation by a constant-current- driven p–n junction: as the measurement time becomes longer, so does the degree of squeezing which is measured as $\langle \Delta n_{\text{e}}^2 \rangle / \langle n_{\text{e}} \rangle$. For a typical LED operating at room temperature, this fundamental noise limit $\langle \Delta n_{\text{e}}^2 \rangle$ is on the order of $10^7 \sim 10^8$. Thus, a quantum optical repeater using a conventional LED cannot operate in a single-photon limit [267]. This fundamental limit of intensity noise squeezing manifests itself by a finite squeezing bandwidth given by [103]

$$B = \frac{1}{2\pi\tau_{te}} = \frac{eI}{2\pi k_B T C_{dep}} \quad . \tag{3.3}$$

There is another source of stochasticity in addition to thermionic emission or tunneling of carriers in a constant-current-driven LED, which is the radiative recombination of the injected carriers. When this is taken into account, the squeezing bandwidth over which the intensity noise is reduced to below the shot-noise value is given by Eq. (2.40):

$$f_{3dB} = \frac{1}{2\pi(\tau_{te} + \tau_{sp})} = \frac{1}{2\pi(\frac{k_B T C_{dep}}{eI} + \tau_{sp})} \quad , \tag{3.4}$$

where τ_{sp} is the radiative lifetime. Therefore, the squeezing bandwidth should be proportional to the current, I, and inversely proportional to the temperature, T, and the capacitance, C_{dep}, in a low-current regime, but is limited by the radiative recombination lifetime in a high- current regime. An experimental effort to observe this squeezing bandwidth dependence and demonstrate this internal regulation mechanism ("macroscopic Coulomb blockade") is presented in the following section.

3.3 Measurement of the Squeezing Bandwidth

Wideband squeezing, of up to 1 GHz, was observed in a constant-current-driven semiconductor laser [162], suggesting that the bandwidth determined using the macroscopic Coulomb blockade effect is beyond this value when the junction capacitance is small and the driving current is large. Preliminary experimental attempts to measure the squeezing bandwidth on a semiconductor laser showed that the squeezing bandwidth was limited by the tail of a large relaxation oscillation peak before the limitation due to macroscopic Coulomb blockade effect sets in. The next experiment was performed using light-emitting diodes (LEDs), where the optical feedback mechanism is absent. Since the ultimate level of squeezing is determined by the quantum efficiency to detect the photons, a high-quantum-efficiency LED is required. The quantum efficiency of a simple GaAs LED is limited to about 2 % (per facet) even if the internal electron-to-photon conversion efficiency is 100 %, due to total internal reflection at the semiconductor–air interface. Such total internal reflection is not an issue for a semiconductor laser, since the photon output is achieved by finite reflectivity of the end mirror for the lasing mode.

High-quantum-efficiency LEDs were fabricated using the molecular beam epitaxy (MBE) growth technique by our collaborators at Hamamatsu Photonics in Japan. The schematic band structure of the LED, with a forward-bias voltage applied to reach a near flat-band condition, is shown in Fig. 3.1. A heavily doped p-type $Al_{0.1}Ga_{0.9}As$ buffer layer was grown on a p-type GaAs substrate. The actual junction consists of an Al-concentration graded p-type $Al_xGa_{1-x}As$ ($0.1 < x < 0.35$) layer and a graded n-type $Al_xGa_{1-x}As$

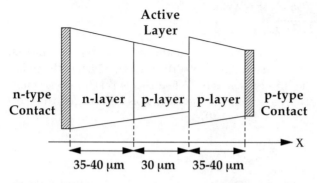

Fig. 3.1. Schematic band structure of the LED used in this experiment, with a forward-bias voltage to reach a near flat-band condition

(0.1 < x < 0.35) layer. Standard ohmic contacts of Ni/Au/Ge and Zn/Au were put down for the n-type top layer and p-type substrate, respectively. The graded Al concentration plays a dual role of photon confinement due to index gradient and photon recycling due to re-absorption. When an electron–hole pair recombines to produce a photon in the active layer, the photon can be either emitted towards the surface or towards the substrate. In a conventional p–n junction, the photons emitted towards the substrate are completely lost and the photons emitted towards the surface have only an about 2 % chance of escaping the GaAs crystal into air (per facet) due to total internal reflection. In such a graded junction, the photons reflected from the n-side top-layer–air interface have a chance of being re-absorbed and re-emitted, increasing the chance of escape from the crystal. Furthermore, the photons emitted towards the substrate will also have the chance to be re-absorbed and re-emitted towards the surface. Such photon recycling increases the overall quantum efficiency of the LED at the sacrifice of an increased effective-carrier lifetime. The carrier lifetime due to spontaneous emission in these LEDs was measured by pumping the LED with a pulsed semiconductor laser ($\sim$5 ns pulse width) and looking at the decay time constant of the spontaneous emission with an avalanche photodiode. The carrier lifetime of the LED was measured to be about $\tau_{\mathrm{sp}} = 260$ ns, which corresponds to a bandwidth of $\sim$610 kHz. Four LEDs with different junction areas (0.073, 0.423, 1.00 and 2.10 mm^2) were fabricated, which provides four different values of junction capacitances.

A schematic of the experimental setup is shown in Fig. 3.2. The LED was coupled in a face-to-face configuration with a high-quantum-efficiency photodetector (Hamamatsu S3994 for room-temperature measurements, and Hamamatsu S1722-01 for lower-temperature measurements). The overall quantum efficiency (photodetector current / LED driving current) was 12$\sim$21 % at room temperature and 8$\sim$10 % at lower temperatures. The lower quantum efficiency at lower temperatures is mainly because the photodetector used in the low temperature (S1722-01) had a smaller area and the

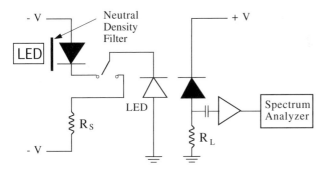

Fig. 3.2. A schematic of the experimental setup to measure the squeezing bandwidth of an LED

collection efficiency was lower. The LEDs were driven either by a high-impedance constant-current source (with 800 Ω series resistance) to generate sub-Poissonian light or by a shot-noise-limited current to generate Poissonian light. A shot-noise-limited current was generated by a photodiode coupled weakly to another LED. This scheme makes it possible to avoid any change in the coupling geometry when calibrating the shot-noise value. The linearity of the photodetector was confirmed up to 40 mA of photocurrent for the case of S3994 detectors. This enables one to drive the LEDs with currents as high as 40 mA and still be able to calibrate the shot noise precisely. The noise current generated from the photodetector was amplified by a homemade ultra-low-noise preamplifier. The signal in this measurement is the photocurrent noise converted to a voltage signal via the load resistor R_L. The signal power is proportional to the square of R_L. The total noise in this measurement is the sum of the thermal noise generated at R_L and the input noise of the amplifier used. The signal-to-noise ratio (SNR) is given by

$$SNR = \frac{2eI_{ph}FR_L^2}{4k_BTR_L + S_{amp}}, \tag{3.5}$$

where I_{ph} is the photocurrent, $F < 1$ characterizes the amount of squeezing, and S_{amp} is the input noise power spectral density of the amplifier. When the amplifier input noise, S_{amp}, is small compared with the thermal noise of the load resistor, the SNR is proportional to the load resistor used. A high impedance front-end design using a 10 kΩ load resistance was used to improve the SNR. The upper limit for the resistor value is set by the 3 dB rolloff bandwidth of the detection circuit, which is determined by the load resistor and the capacitance of the photodetector. With a 10 kΩ load resistor, the rolloff was about 5 MHz for S3994 detectors and 30 MHz for S1722-01 detectors. These detector bandwidths are well above the measurement frequency ($\leq$1.5 MHz). The amplified noise current is finally fed into a spectrum analyzer (HP70908A), which is connected to a personal computer for data

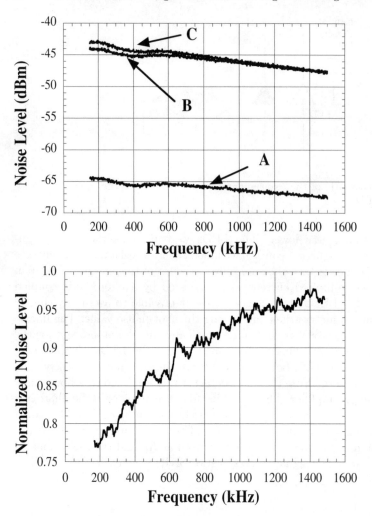

Fig. 3.3 a, b. A typical set of noise measurement data. The photocurrent was 4.73 mA, and the temperature was 295 K. (**a**) The noise spectra measured by the spectrum analyzer. Trace A is the thermal background noise, trace B is the photocurrent noise when the LED was driven with a constant-current source, and trace C is the photocurrent noise when the LED was driven with a shot-noise-limited current source. (**b**) The intensity noise of the constant-current-driven LED normalized by the shot-noise value. The thermal background noise was subtracted from both traces B and C before normalization

analysis. The whole setup including the LEDs, the photodetectors and the preamplifier circuit, was mounted in a cryostat, and cooled to 77 K.

Figure 3.3 shows a typical set of noise measurement data. Figure 3.3a is the data taken directly from the spectrum analyzer. For this specific case, the photocurrent was 4.73 mA, and the overall quantum efficiency was ~20 %

(the differential quantum efficiency was ~22 %). Traces A, B and C show the thermal background noise, the photocurrent noise when the LED was driven with a high-impedance constant- current source, and the photocurrent noise when the LED was driven with a shot-noise-limited current source, respectively. The photocurrent noise was about 20 dB above the thermal noise, and the detector response was reasonably flat in the frequency region of interest. One can see that the photocurrent noise for a constant-current-driven LED is below the shot-noise value in the low-frequency regime, but it approaches the shot-noise value in the higher- frequency regime. The thermal noise trace (A) was subtracted from the two photocurrent noise traces (B and C), and the squeezed noise trace (B) was normalized by the shot-noise-limited trace (C). Figure 3.3b shows such a normalized noise spectrum. The maximum squeezing observed in the low-frequency region is about 0.21 (1.0 dB), which is in good agreement with the expected value from the overall quantum efficiency. The squeezing bandwidth was determined to be the frequency at which the degree of squeezing is reduced by a factor of two and was found to be ~720 kHz in this specific case.

Similar measurements and analyses were performed for the LEDs with different junction areas, as functions of the driving current (1~40 mA) and operation temperature (78~295 K). Figure 3.4a–d shows the squeezing bandwidth of four LEDs at room temperature. The dotted lines are the expected squeezing bandwidths due to macroscopic Coulomb blockade effect (3.3). The dashed lines show the lifetime limitation of the squeezing bandwidth, with $\tau_{rad} = 290$ ns and the corresponding bandwidth of 560 kHz, that best fits the data. This value is in good agreement with the independent measurement performed by the optical excitation method. The solid lines show the theoretical squeezing bandwidth according to (3.4). The only fitting parameters in the curves are the spontaneous emission lifetime and the junction capacitance C_{dep}. In a low-current regime, the squeezing bandwidth increases linearly with increasing the current. The four capacitance values used to fit the measurement curves were 6.5 nF, 30 nF, 90 nF and 180 nF. The ratios of these four capacitance values are 0.072 : 0.33 : 1.0 : 2.0. They are in close agreement with the ratios of the junction areas, 0.073 : 0.42 : 1.0 : 2.1. An independent measurement confirmed that the junction capacitance is proportional to the junction area. Therefore, we conclude that the squeezing bandwidth is proportional to the current and inversely proportional to the capacitance in the low-current regime.

As the driving current increases, the squeezing bandwidth is limited by the carrier recombination lifetime and saturates at ~560 kHz. For the smallest-area LED, the squeezing bandwidth increases above this value at higher driving currents. This is attributed to the carrier-concentration-dependent radiative lifetime. At higher current densities, the carrier density in the active region increases, and the carrier lifetime is shortened [100]. This results in the increased squeezing bandwidth.

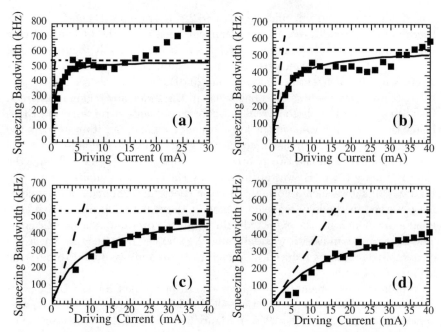

Fig. 3.4 a–d. The squeezing bandwidth as a function of a driving current for various LEDs at room temperature. *Dashed lines* show the radiative recombination lifetime limitation of about 560 kHz (carrier lifetime of 290 ns). *Dotted lines* are the squeezing bandwidths expected from (3.3). *Solid lines* are the expected overall squeezing bandwidth expressed by (3.4). Areas of the LEDs (capacitance values to fit the data) are (**a**) 0.073 mm^2 (6.5 nF), (**b**) 0.423 mm^2 (30 nF), (**c**) 1.00 mm^2 (90 nF) and (**d**) 2.10 mm^2 (180 nF)

Figure 3.5a–d shows the squeezing bandwidth of the LED with an area of 1.00 mm^2 measured at different temperatures. The squeezing bandwidth was measured at 295 K, 220 K, 120 K and 78 K. Again, the dotted lines show the squeezing bandwidth limitation due to the macroscopic Coulomb blockade effect, and the dashed lines show the limitation due to the recombination lifetime at ∼560 kHz. The solid lines are the theoretical squeezing bandwidth limitation expected by (3.4). There are no fitting parameters in the curves, except for the capacitance $C_{dep} = 90$ nF obtained by fitting the data in Fig. 3.4c. Actual temperatures at which the measurements were made were used to draw the curves. From this data, one can see that the squeezing bandwidth is linearly proportional to the driving current in the low-current regime and saturates at the radiative lifetime limited value of ∼560 kHz in the high-current regime. The linear slope in the low-current regime is inversely proportional to temperature. Close agreement between the experimental result and the simple theoretical model described by (3.4) can be seen.

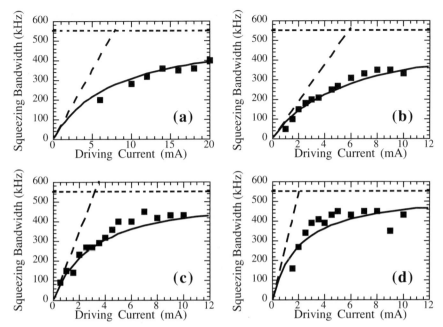

Fig. 3.5 a–d. The squeezing bandwidth as a function of driving current for various temperatures. The area of the LED was 1.00 mm². The temperatures are (**a**) 295 K (identical to Fig. 3.4c), (**b**) 220 K, (**c**) 120 K and (**d**) 78 K. *Dashed lines* show the radiative recombination lifetime limitation of about 560 kHz. *Dotted lines* are squeezing bandwidths expected from (3.3). *Solid lines* are the overall bandwidths expected from (3.4). A junction capacitance of 90 nF and a carrier lifetime of 290 ns obtained by room-temperature measurement (Fig. 3.4) were used

It was observed some time ago that the depletion layer capacitance, C_{dep}, gives rise to a cutoff frequency in the intensity modulation bandwidth other than the carrier lifetime in semiconductor lasers and LEDs [88, 100, 147]. However, the cutoff frequency in these classical modulation experiments originates from the time constant defined by the junction capacitance (C_{dep}) and the external source resistance (R_{s}) in the modulation circuit, $R_{\mathrm{s}}C_{\mathrm{dep}}$. The experimental result reported here is fundamentally different. What has been investigated here is the internal self-regulation mechanism for electron injection across the junction depletion layer, which is totally independent of the $R_{\mathrm{s}}C_{\mathrm{dep}}$ time constant in the external circuit.

3.4 Summary

The squeezing bandwidth of a constant-current-driven p–n junction LED was measured as a function of current, capacitance and temperature. The squeezing bandwidth in a low current regime is (a) linearly proportional to the driving current, (b) inversely proportional to the junction capacitance, and (c) inversely proportional to the temperature. These experimental results provide strong evidence for the theoretical predictions that the macroscopic Coulomb blockade effect is the fundamental mechanism for intensity squeezing in a constant-current-driven p–n junction [103, 123].

4. Amplitude-Squeezed Light Generation in Semiconductor Lasers

4.1 Introduction

In Chap. 2 and 3, we discussed the principles of pump-noise suppression in constant-current-driven p–n junction light-emitting devices. Such a pump-noise suppression mechanism provides one necessary condition for the generation of squeezed states in semiconductor lasers, and sub-Poissonian light in light-emitting diodes (LEDs).

Although the injection of carriers into the active region of p–n junction light-emitting devices is perfectly regulated, there can be other random processes before the injected carriers are converted into photons and detected at a photodetector. One example of these in an LED is the radiative recombination process, discussed in Chap. 2 and 3. The story is more complicated in the case of a semiconductor laser, where the nonlinear laser oscillation process takes place.

It is generally believed that the intensity noise of the photons generated by a semiconductor laser should reflect the pump noise when the laser is driven far above the oscillation threshold [161]. Under such a driving condition, a highly saturated gain guarantees that all the pumped carriers add photons to the main mode of the laser output, and therefore potential noise contributions from the nonlinear processes in the laser are minimal. Therefore, constant-current driving (to suppress pump noise) and high pumping (to achieve high gain saturation) were considered to be the two conditions necessary for amplitude-squeezed state generation in a semiconductor laser.

Several different types of semiconductor lasers were tested after the initial experimental demonstrations of amplitude-squeezed states [161, 198] and it was found that the measured intensity noise was often above the shot-noise level, even if the two requirements (constant-current source and high pumping) were satisfied. This fact suggested that some other requirements had to be satisfied in order to generate amplitude-squeezed states.

The side-mode intensities of a typical semiconductor laser are only 20 to 30 dB below the main-mode intensity. These relatively large side modes compete with the main-mode for the excited carriers, which introduces huge fluctuations to the main mode intensity. These fluctuations are called longitudinal-mode-partition noise. When the gain medium of a semiconductor laser is homogeneously broadened, it is the same population of carriers which emits in all the modes. Consequently, negative correlation occurs between the intensity noise of the main mode and that of the side modes. Due to this negative intensity noise correlation, the longitudinal-mode-partition noise can often be canceled out in the total intensity noise. Actually, an amplitude-squeezed state has been generated from a multi-longitudinal-mode semiconductor laser even in the presence of very high longitudinal-mode-partition noise. The existence of the negative intensity noise correlation between the main mode and the side modes was also directly observed [105].

However, when the negative intensity noise correlation is imperfect it cannot cancel out the longitudinal-mode-partition noise completely from the total intensity (including all the longitudinal modes) noise. As a result, amplitude squeezing would be degraded by the presence of the residual longitudinal-mode-partition noise in the total intensity noise. This residual longitudinal-mode-partition noise is considered to be one of the reasons why some semiconductor lasers do not generate amplitude-squeezed states even when their quantum efficiencies (QEs) are very high and they are pumped well above the threshold by a constant-current source. The issue of longitudinal-mode-partition noise and its impact on amplitude-squeezed state generation is the central topic of Chap. 4 and 5.

It is well known that external feedback and injection-locking can suppress the side-mode intensities. In some cases, the external-cavity-controlled [63] and injection-locked [106, 242] semiconductor lasers can generate amplitude-squeezed states even when the intensity noise of the free-running lasers is above the shot-noise level. We give the following qualitative explanation for the intensity noise reduction.

The longitudinal-mode-partition noise is suppressed by external self-feedback or injection-locking because of the side-mode intensity suppression. The residual longitudinal-mode-partition noise which cannot be suppressed by the imperfect negative intensity noise correlation would be much smaller than the intensity noise of the free-running condition. Therefore, the degradation of amplitude squeezing due to the residual longitudinal-mode-partition noise is avoided, and the intrinsic amplitude squeezing is observed.

The quantitative relation between longitudinal-mode-partition noise and amplitude squeezing had not been investigated beyond the above-mentioned qualitative argument until recently. We have performed measurements of the longitudinal-mode-partition noise of free-running, grating-feedback external-cavity and injection-locked semiconductor lasers [109]. We used a Mach–Zehnder interferometer with unequal arm-lengths (nonsymmetric) to separate the longitudinal modes so that the longitudinal-mode-partition noise does not cancel out.

4.2 Interferometric Measurement of Longitudinal-Mode-Partition Noise

4.2.1 Principle

The longitudinal-mode-partition noise has been measured by discriminating the main mode from the other side modes using a grating monochrometer [105, 165, 276]. However, this method cannot give us the exact magnitude of the longitudinal-mode-partition noise because of the high insertion loss of the monochrometer. A more precise measurement of the longitudinal-mode-partition noise can be performed using a nonsymmetric interferometer [108].

The longitudinal mode separation Δf of a typical semiconductor laser is about 100 GHz, which is much smaller than the center wavelength (on the order of 1 μm, or 10^5 GHz). After traveling through the two arms of an interferometer, the accumulated phase difference between the two beams is $\Delta \phi = 2\pi \Delta l / \lambda$, where Δl is the arm-length difference and λ is the wavelength. Therefore, when one arm-length is changed by a few wavelengths around $\Delta l = 0$, $\Delta \phi$ for different longitudinal modes is almost identical. Thus, the fringe patterns of all the modes of interest (about 100 side modes around the main lasing mode) overlap very well, as shown in Fig. 4.1a. However, when Δl is increased, $\Delta \phi$ for different longitudinal modes begins to differ significantly. For $\Delta l_0 \equiv c/(2\Delta f)$, where c is the speed of light, $\Delta \phi$ differs by π for adjacent longitudinal modes. This means that if the interferometer phase difference for the main mode is $\Delta \phi_0$, then the phase difference for either of the first side modes is $\Delta \phi_0 \pm \pi$ and that for the nth side modes is $\Delta \phi_0 \pm n\pi$. Consequently, at the bright fringe of the main mode ($\Delta \phi_0 = 2k\pi$, where k is an integer), all of the even-numbered side modes interfere constructively, while all of the odd-numbered side modes interfere destructively, as shown in Fig. 4.1b. The opposite is true at the dark fringe. Similarly, for $\Delta l = 2\Delta l_0$, $\Delta \phi$ for adjacent modes differs by 2π; thus, the fringes of all the modes overlap again. Therefore, by setting the arm-length difference Δl to $c/(2\Delta f)$, we can discriminate between longitudinal modes and measure the longitudinal-mode-partition noise with the highest efficiency.

4.2.2 Experimental Setup

Figure 4.2 shows the experimental setup for measuring the longitudinal-mode-partition noise of semiconductor lasers using a Mach–Zehnder interferometer. The laser output was injected into a Mach–Zehnder interferometer with the arm-length difference $\Delta l = c/(2\Delta f)$. One of the outputs from the interferometer was used to monitor the interference fringe. The other output was divided into two beams, using a half-wave plate and a polarizing beam splitter, and detected by large-area photodiodes (Hamamatsu S3994). The output signals from the photodetectors were amplified and fed into a π-hybrid mixer. The sum signal (the intensity noise of the interferometer

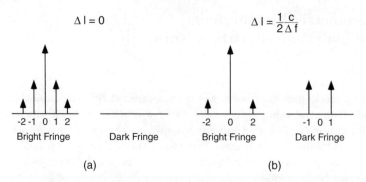

Fig. 4.1 a, b. Schematic spectra of a bright fringe and a dark fringe for the arm-length differences (**a**) $\Delta l = 0$ and (**b**) $\Delta l = c/(2\Delta f)$

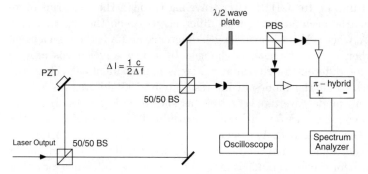

Fig. 4.2. The experimental setup for measuring the longitudinal-mode-partition noise of semiconductor lasers. BS: beam splitters; PBS: polarizing beam splitter; and PZT: piezoelectric transducer

output) and the difference signal (the corresponding shot-noise value) were measured using a radio-frequency (RF) spectrum analyzer. We also measured the intensity noise of the total modes and the corresponding shot-noise value bypassing the interferometer.

Using this setup, we measured the mode-partition noise of semiconductor lasers in different configurations, in which the intensities of the side modes were varied. These measurements provided an understanding of the quantitative relation between the side-mode intensity (side-mode suppression ratio, SMSR) and the residual longitudinal-mode-partition noise which degrades the squeezing.

4.3 Grating-Feedback External-Cavity Semiconductor Laser

In this section, we discuss the contribution of the longitudinal-mode-partition noise to the total intensity noise of a laser in both a free-running and an external-cavity configuration [109].

4.3.1 Experimental Setup and Procedure

We used a single-quantum-well (SQW) semiconductor laser (SDL 5401) in the grating-feedback external-cavity configuration. The cavity length of the semiconductor laser was 750 μm and the longitudinal-mode separation Δf was about 50 GHz. The SMSR of the free-running laser was about 20 dB. A gold-coated holographic grating with 1200 lines/mm was placed 3 cm away from the laser. The first-order diffraction (with 15% efficiency) was fed back into the laser and the zeroth order (with about 84% efficiency) output from the cavity was used in the experiment. The laser temperature was actively stabilized near 25°C and the threshold current of the external-cavity laser was 15 mA.

We measured the intensity noise at one of the output ports of the interferometer as we changed the pumping level in both the free-running and the external-cavity configuration. First, the arm-length difference was set to zero ($\Delta l = 0$). The intensity noise at a bright fringe (which reflected the total laser intensity noise, say S_{L}) and the corresponding shot-noise value (say S_{S}) were measured. The fringe visibility was about 95%. The optical loss from the laser to the detectors was −1.5 dB. The intensity noise at the dark fringe was from leakage of the main mode because of imperfect fringe contrast. The shot-noise level corresponding to the dark fringe was 0.5 dB above the thermal noise floor.

We then changed the length of one arm such that $\Delta l = c/(2\Delta f) = 3$ mm. At a bright fringe the main mode and all even side modes interfered constructively, while all odd side modes interfered destructively, and vice versa for a dark fringe. Again, the intensity noise at a bright (say S_{B}) and a dark (say S_{D}) fringe was measured. We checked that S_{D} for the external-cavity laser was larger than what was expected from the leaking main-mode power; thus, it was indeed the intensity noise of the odd side modes. The measured absolute QE after correcting for losses outside the laser was 0.38. This value included the loss from the grating (1%), collimating lens, and rear facet, as well as the laser internal losses.

4.3.2 Experimental Results

Figure 4.3 plots the noise of all even modes (main mode together with the even side modes), all odd modes and total intensity (all modes) for the free-running

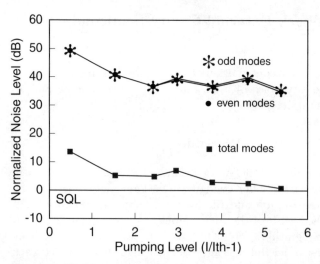

Fig. 4.3. The intensity noise of even modes, odd modes and total intensity for the free-running laser. All curves are normalized by the shot-noise for the total power. Experimental data is corrected for the loss in the setup. The measurement frequency was 10 MHz and resolution band width was 300 kHz

laser. All three noise values are normalized by the shot noise, corresponding to the total power. The noises of the even modes (measured at the bright fringe, S_B) and the noise of the odd modes (measured at the dark fringe, S_D) were almost equal and were about 30–35 dB larger than the total intensity noise, S_L. Note that power in all the odd modes was about 20 dB smaller than the power in the even modes, yet the noise of the odd modes was equal to the noise of the even modes. The noise in the even modes above the total intensity noise is the longitudinal-mode-partition noise for these modes.

The very high longitudinal-mode-partition noise was suppressed by the negative intensity noise correlation between the even modes and the odd modes when all the modes were detected together. From the magnitude of the suppression of the longitudinal-mode-partition noise, we estimated the strength of the negative intensity noise correlation between these two sets of modes to be more than 99.96%. The anticorrelated noise between the main mode and all the even side modes is cancelled out when all of these modes are detected together, so it does not contribute to S_B. The longitudinal-mode-partition noise of the main mode is approximately equally contributed to by even and odd side modes when a large number (about 100) of side modes participate in the total intensity noise. Thus, S_B includes half of the longitudinal-mode-partition noise of the main mode alone.

Figure 4.4 plots the noise of all even modes (main mode together with the even side modes), all odd modes and total intensity (all modes) for the external-cavity laser. All three noise values are normalized by the shot noise corresponding to the total power. In this case, the difference between the even

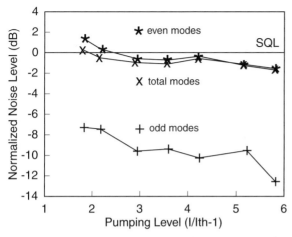

Fig. 4.4. The intensity noise of even modes, odd modes and total intensity for the external-cavity laser. All curves are normalized by the shot-noise for the total power. Experimental data is corrected for the loss in the setup. The measurement frequency was 10 MHz and resolution band width was 300 kHz

modes (S_B) and the total intensity (S_L) noise was only 1 dB at a pumping level of 3 times the threshold and it became almost zero near 6 times the threshold. S_L was squeezed when the pumping level was more than 3 times the threshold. The removal of the odd modes did not enhance the intensity noise in the even modes. This strongly suggested that the longitudinal-mode-partition noise was negligible compared to the total intensity noise. In this case the dark-fringe noise S_D provided a measure of the small but nonzero longitudinal-mode-partition noise. At high pumping levels, S_D was about 12 dB below S_B and 3–4 dB above the shot-noise level corresponding to the dark-fringe intensity. This implied that the odd modes had excess intensity noise, while the main mode was squeezed.

We then measured total intensity noise (bypassing the interferometer) of the free-running laser and the external-cavity laser. We compared measurements, after correcting for loss, with the theoretical value calculated from the observed QE. The intensity noise of the free-running laser was a minimum of 2 dB above the shot-noise level for pumping levels near 6 times the threshold. On the other hand, the intensity noise of the external-cavity laser showed −1.5 dB of squeezing (−0.9 dB directly measured) at similar pumping levels, which is in good agreement with the single-mode theory.

In order to determine the cause of the excess noise, we measured the polarization-partition noise. Introducing a polarizing beam splitter to reject the weak polarization (transverse magnetic, TM) did not influence the intensity noise of the external-cavity laser and slightly increased the intensity noise of the free-running laser. The polarization-partition noise did not contribute significantly to the observed excess noise.

4.3.3 Discussion

From these experimental results, we conclude that the excess noise of the free-running laser was due to the residual longitudinal-mode-partition noise, which was not completely suppressed because of the imperfect negative intensity noise correlation among the longitudinal modes. At high pumping levels of the external-cavity laser, the longitudinal-mode-partition noise was negligible compared to the intensity noise of the main mode. Therefore, the imperfect negative intensity noise correlation did not enhance the total intensity noise significantly. However, when the longitudinal-mode-partition noise was large, the imperfect negative intensity noise correlation increased the intensity noise as we saw in the case of the free-running laser. The difference between the intensity noises of the free-running laser and the external-cavity laser (at high pumping level) is thus a measure of the residual longitudinal-mode-partition noise which exists in the total intensity noise.

4.4 Injection-Locked Semiconductor Laser

In this section, we will discuss the relation between amplitude squeezing and the suppression of the longitudinal-mode-partition noise by injection-locking [109].

4.4.1 Experimental Setup and Procedure

We used a free-running single-QW semiconductor laser (SDL 5401) as a master laser for injection-locking. The master laser was mounted in a temperature-controlled chamber and its temperature was actively stabilized near 25°C with an accuracy and stability of several millikelvin. At this temperature, the threshold current was 18 mA and the operating current was around 77 mA. The master laser oscillated in a single longitudinal mode ($\lambda = 787$ nm), with side-mode intensities approximately 20 dB below the main-mode intensity. The slave laser was a transverse-junction-stripe (TJS) semiconductor laser (Mitsubishi ML3401) with an antireflection coating ($\simeq 10\%$) on the front facet and a high-reflection coating ($\simeq 90\%$) on the rear facet. This slave laser was mounted in a cryostat and cooled down to approximately 107 K. The threshold current was 1.5 mA and the operating current was 28.66 mA. The side-mode intensity of the free-running slave laser was about 13 dB below the main-mode intensity. The longitudinal-mode separation of the slave laser was about 150 GHz; therefore, we set the arm-length difference Δl to $c/(2\Delta f) = 1$ mm. The visibility of the interference was 97.4% with the zero arm-length difference and 78.5% with the arm-length difference of 1 mm. This decrease of visibility with the arm-length difference of 1 mm is due to the destructive interference of the odd side modes at a bright fringe and their constructive interference at a dark fringe.

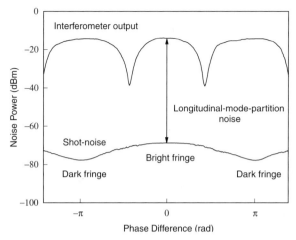

Fig. 4.5. The intensity noise variation of the interferometer output as one arm-length is changed by about one wavelength. The arm-length difference of the Mach–Zehnder interferometer was 1 mm. The measurement frequency was 20 MHz. The *lower curve* shows the variation of the corresponding shot-noise value

Figure 4.5 shows the intensity noise and the corresponding shot-noise variation in the output of the interferometer for an arm-length difference of 1 mm, as one arm-length is changed by about one wavelength (the traces are explained in detail in [108]). Here, the free-running slave laser output was injected into the interferometer. The measurement frequency was 20 MHz. The total intensity noise of the free-running slave laser output was 1.2 dB below the shot-noise level. The intensity noise of even modes (the main mode and all the even side modes) was 55 dB above the shot-noise level. Since the even mode intensity noise was a few orders of magnitude higher than the total intensity noise, it was almost completely contributed by the longitudinal-mode-partition noise. Thus, we will refer to the even mode noise as the longitudinal-mode-partition noise. The measurements indicated that the very high longitudinal-mode-partition noise was reduced below the total intensity noise by negative intensity noise correlation among the longitudinal modes. From the magnitude of the suppression of the longitudinal-mode-partition noise, we estimated the strength of the negative intensity noise correlation to be more than 99.9997%.

4.4.2 Experimental Results

Figure 4.6 shows the longitudinal-mode-partition noise as a function of the locking bandwidth. The longitudinal-mode-partition noise was normalized by the shot-noise level corresponding to the total power. The locking bandwidth was changed by changing the injection power into the slave laser. It is proportional to the square root of the injected power. Here, the longitudinal-mode-

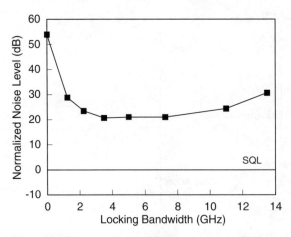

Fig. 4.6. The longitudinal-mode-partition noise as a function of the locking bandwidth. The longitudinal-mode-partition noise is normalized by the shot-noise level corresponding to the total power before interferometer. The measurement frequency was 20 MHz. SQL: standard quantum limit

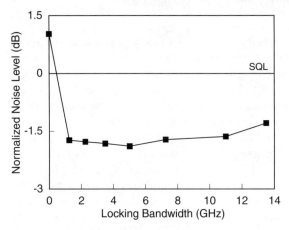

Fig. 4.7. The total intensity noise as a function of the locking bandwidth. The total intensity noise is normalized by the shot-noise level corresponding to the total power before interferometer. The measurement frequency was 20 MHz

partition noise at zero locking bandwidth is the longitudinal-mode-partition noise of the free-running slave laser output. The longitudinal-mode-partition noise decreased as the locking bandwidth was increased up to about 5 GHz. The minimum longitudinal-mode-partition noise was about 21 dB above the shot-noise level. As the locking bandwidth exceeded 5 GHz, the longitudinal-mode-partition noise started increasing.

Figure 4.7 shows the total intensity noise as a function of the locking bandwidth. We chose the driving current at which the intensity noise of the free-running slave laser output was excess noise (about 1 dB) in order to see the difference between the longitudinal-mode-partition noise of the squeezed output (Fig. 4.6) and that of the excess noise output. It turned out that the magnitude of the longitudinal-mode-partition noise of the excess noise output was almost the same as that of the squeezed output. This suggested that by changing the driving current the negative intensity noise correlation among the modes was decreased. It resulted in an about 2 dB increase of the total intensity noise. The total intensity noise decreased as the locking bandwidth was increased. The maximum amplitude squeezing (-1.9 dB) was observed when the locking bandwidth was 5 GHz (the suppression of the longitudinal-mode-partition noise was 34 dB compared to the free-running case). The amplitude squeezing was degraded by the further increase of the locking bandwidth. This degradation of the amplitude squeezing corresponds to the increase of the longitudinal-mode-partition noise in Fig. 4.6.

4.4.3 Discussion

These experimental results show two important points. First, as expected, the intensity noise and the longitudinal-mode-partition noise were both reduced by the injection-locking, which decreased the side-mode intensities. Second, for the injection-locked laser, both noises showed similar dependence on the injection power or the locking bandwidth. The longitudinal-mode-partition noise gradually increased when the injection power was increased beyond an optimal value. The intensity noise also increased accordingly.

These observations strongly support the argument that the excess noise of the free-running laser, as well as the injection-locked laser at the high injection power, was due to the residual longitudinal-mode-partition noise. The longitudinal-mode-partition noise was not completely suppressed because of the imperfect negative intensity noise correlation among the longitudinal modes. The intensity noise squeezing of injection-locked laser at optimal injection power was close to what is expected from the measured QE.

Note that the longitudinal-mode-partition noise of the injection-locked laser was a minimum of 20 dB above the shot noise, whereas the longitudinal-mode-partition noise of the external-cavity lasers was negligible. This implies that the external cavity suppresses the side modes more effectively. The reason for this difference is not clear.

The gradual increase in the longitudinal-mode-partition noise of the injection-locked laser at the higher injection power indicated a small increase in the size of the side modes. We did not observe a change in the spectrum recorded by an optical spectrum analyzer when the injection power increased beyond the optimal value. However, the small change in the size of side modes could not have been observed directly. This is because the resolution limit of the spectrum analyzer was 0.1 nm, which was very close to the separation of side modes, 0.33 nm. Moreover, the side modes were more than 20 dB smaller than the main mode.

Note that the parallel enhancement of the total intensity noise with the increase in the longitudinal-mode-partition noise does not only come from the larger longitudinal-mode-partition noise, but also from any further degradation in the correlation among the longitudinal modes caused by the injected power.

It is known that the injection-locking reduces the gain saturation in the slave laser and modifies the gain profile or gain spectrum. This modification might alter the side-mode intensities. The simple theories of injection-locking assume only a single longitudinal mode for both the slave and the master laser [75, 89]. These theories consider suppression of the gain at a single wavelength, not of the whole gain spectrum. Therefore, the existing theories cannot explain our observation on a multi-longitudinal-mode laser.

The theory developed by Gillner et al. [75] predicts that beyond a certain value the high injection power enhances the intensity noise because of its own shot noise. However, the injection power in our experiments was well below this limit. We need a theory which considers multi-longitudinal-mode operation of the slave laser to explain our results. We developed a model for the modification of the gain spectrum and performed a calculation of the noise of a multi-longitudinal-mode injection-locked laser to explain the dependence of the longitudinal-mode-partition noise on the injection power.

4.4.4 Modeling of the Noise of an Injection-Locked Laser

The injected power at a particular wavelength reduces the gain required for the lasing at that wavelength. This reduction in the gain of the injection-locked mode decreases the gain at all frequencies within the homogeneous linewidth because of the intraband carrier scattering. The homogeneous linewidth of semiconductor lasers is several tens of nanometers. However, the carrier scattering does not simply scale down the gain, instead it makes the gain smaller and broader.

In addition, because of the large anomalous dispersion in the semiconductor gain medium (represented by the parameter α) [95], change in the carrier density also changes the refractive index. Because of this dispersion, the gain on the blue (lower wavelength) side of the gain peak is suppressed differently compared to the red (higher wavelength) side. Consequently, the shape of gain spectrum changes, and it becomes asymmetric.

Bouyer et al. [30] studied injection-locked InGaAsP lasers both experimentally and theoretically. They observed a few different regimes of the operation of the injection-locked slave laser. In the different regimes, the dependence of the spectrum on the injection power was different. We will not go into the details of their study here, but we will briefly mention the important results. Their experiments on multi-longitudinal-mode lasers showed that, when the injection-power was increased above a critical value, the side modes were not suppressed any further. In this saturation regime, as expected, the locking bandwidth increased proportional to the square root of the injected power. They noted two such saturation regimes when the injection power was varied in a large range. They attributed this behavior to the asymmetry of the gain spectrum and the shift of the gain maximum towards higher wavelengths.

In our experiment the increase in the longitudinal-mode-partition noise at higher injection power implies an increase in the size of the side modes rather than the saturation. Since we were measuring noise, we could measure small changes in the spectrum more precisely. As mentioned earlier, we also did not observe any change in the spectrum recorded by a spectrum analyzer.

We included modifications of gain in the Langevin rate equation model for an injection-locked laser with multi-longitudinal modes. Gillner et al. [75] developed the Langevin rate equation model for the noise analysis of a single-mode injection-locked laser. Inoue et al. [105] proposed a rate equation model for a free-running multi-longitudinal mode laser. They considered three longitudinal modes. We combined the two models with a change in the expression for the gain.

The rate equation model for multi-longitudinal modes will be discussed in the next chapter. Here we will describe a model for the dependence of the gain on the injection power and its effect on the noise. In our model, we considered three modes, the main mode and two symmetrically placed first-order side modes. These modes will be denoted by the suffixes 0 and ± 1, respectively. The wavelength of the injected power was taken to be the same as that of the main mode, i.e., the main mode was locked by the injection.

On the basis of the experimental observations, we assumed the following expression for the gain of the side modes:

$$G_i = m_i N_c(P_{\text{in}}) + \kappa_i P_{\text{in}} \tag{4.1}$$

where the index i denotes ith longitudinal mode, G_i is the gain under injection-locking, m_i is the gain coefficient in the free-running condition, N_c is the carrier density, P_{in} is the injected power, and κ_i is the gain-broadening factor. N_c determines the linear gain for all the modes. P_{in} determines the locking bandwidth. The suppression of N_c by the injected power represents the uniform reduction in the gain spectrum. In order to model the gain broadening by the injection-locking, we assumed small positive values for the factors $\kappa_{\pm 1}$. The gain asymmetry can be included by making the broadening factor for the upper side mode, κ_{+1}, smaller than the broadening factor for the lower side mode, κ_{-1}.

These two factors were the only fitting parameters in our model. All other parameters, namely, linear gain, loss, spontaneous-emission lifetime and coupling factor, and pumping level were taken from the experiments. Figures 4.8 and 4.9 plot the calculated noise of only the main mode (which is largely the longitudinal-mode-partition noise) and all the modes (the total intensity noise), respectively, as a function of the locking bandwidth. In this calculation κ_{+1} and κ_{-1} were both 7.3×10^{-5}. The corresponding spectrum showed

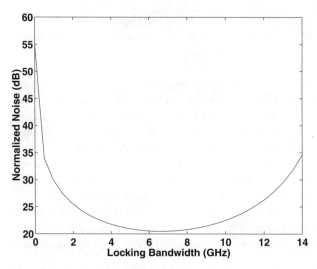

Fig. 4.8. The main-mode noise (longitudinal-mode-partition noise) as a function of the locking bandwidth. The main-mode noise is normalized by the shot-noise level corresponding to the total power

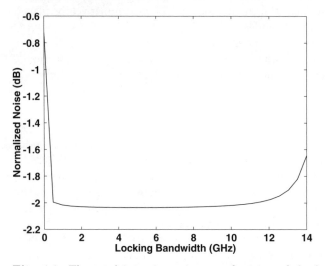

Fig. 4.9. The total intensity noise as a function of the locking bandwidth. The total intensity noise is normalized by the shot-noise level corresponding to the total power

a negligible increase in the size of side modes at a higher level of injection power. The calculated noises match closely with the experimental data shown in Figs. 4.6 and 4.7.

These calculations suggest that, although the locking bandwidth increases with the increase in the injection power the suppression of the side modes becomes slightly less effective at the higher injection power. This result is not expected if one makes the simple assumption that injection-locking uniformly reduces the gain spectrum. The agreement between the model and the experimental observation suggests that it is necessary to consider modification of the gain spectrum under injection-locking. It is the broadening (and asymmetry) of the gain which leads to the increase in the longitudinal-mode-partition noise.

The enhancement of both the longitudinal-mode-partition noise and the intensity noise predicted by the model strongly depends on the gain-broadening factors, $\kappa_{\pm 1}$. We chose the values of $\kappa_{\pm 1}$ to obtain the best fit with the experimental data. It was not possible to get exact values from the known experimental parameters. These parameters assign the effect of the whole spectrum to the two side modes. Therefore, the model is overestimating the change in each individual side mode. However, the model provides insight into the behavior of the multi-longitudinal-mode injection-locked laser.

4.5 Summary

We measured the longitudinal-mode-partition noise of a free-running, grating-feedback external-cavity laser and an injection-locked semiconductor lasers using a nonsymmetric Mach–Zehnder interferometer. Our Mach–Zehnder interferometer provided a sensitive measure of the longitudinal-mode-partition noise by virtue of its low loss and greater than 95% visibility. The experimental results are summarized in Table 4.1, in which the noise level is normalized by the shot-noise level corresponding to the total power.

We observed that the longitudinal-mode-partition noise was much smaller in the external-cavity semiconductor laser than in the free-running semiconductor laser. It became negligible for the external-cavity semiconductor laser at a high pumping level (6 times the threshold). The suppression of side modes by external feedback reduced the residual longitudinal-mode-partition noise arising from the imperfect anticorrelation among different longitudinal modes. Consequently, the intensity noise of the external-cavity laser was squeezed. The measured intensity noise of the external-cavity laser pumped

Table 4.1. Summary of measurements: the longitudinal-mode-partition noise and the total intensity noise versus the side-mode-suppression ratio (SMSR). QW: quantum-well; TJS: transverse-junction-stripe

Laser type	Mode-partition noise (dB)	Total intensity noise (dB)	Side-mode suppression
Free-running QW	$\sim$35 dB	$\geq$2.0 dB	$\sim$20 dB
External-cavity QW	Negligible	-0.9 dB	$\geq$35 dB
Free-running TJS	$\sim$55 dB	-1.2 dB	$\sim$13 dB
Injection-locked TJS	$\sim$21 dB	-1.9 dB	$\sim$25 dB

far above threshold agreed well with the standard single-mode theory for semiconductor laser intensity noise. We did not observe polarization-partition noise in the external-cavity laser. These observations suggested that the excess noise in the free-running laser (operating at high pumping levels) was primarily the residual longitudinal-mode-partition noise. The difference in noise levels for the two is thus a measure of the residual longitudinal-mode-partition noise in the free-running laser.

We also observed that the free-running TJS semiconductor laser which showed -1.2 dB of squeezing had 55 dB of longitudinal-mode-partition noise. This indicated that a very strong negative intensity noise correlation (more than 99.9997%) was formed among the longitudinal modes of the TJS laser. This huge longitudinal-mode-partition noise was suppressed by 34 dB with the optimal locking bandwidth (5 GHz), but it was still 21 dB above the shot-noise level. The further increase in the locking bandwidth gradually increased the longitudinal-mode-partition noise as well as the total intensity noise. The increase in the noise is a result of the broadening and asymmetry of the gain spectrum. The gain spectrum is modified by injection-locking because of the coupling of the carrier density and the refractive index in semiconductor lasers.

We calculated the noise of an injection-locked multi-longitudinal-mode laser using a simple model for the dependence of the gain spectrum on the injection power. In the calculations we included three longitudinal modes. By treating the gain broadening at these modes as free parameters, we could obtain good fit with the experiments.

5. Excess Intensity Noise
of a Semiconductor Laser
with Nonlinear Gain and Loss

5.1 Introduction

In Chap. 4 we discussed that in many semiconductor lasers the residual mode-partition noise may exist in the total output, which enhances the total intensity noise. The residual mode-partition noise arises from the imperfect negative correlation among longitudinal modes. Consequently, squeezing may be lost or impaired even when both conditions required in the original theory – a quiet pump and operation far above the threshold – were satisfied.

It was suggested that it is the inhomogeneity and the nonlinearity of the semiconductor gain medium which result in the excess noise when weak side modes are present [242], since these weak side modes carry a large amount of noise due to amplified spontaneous emission. To represent the inhomogeneity of the gain medium, Marin et al. [165] introduced a fitting term in the Langevin rate equation model. The fitting term caused small self-saturation of each mode by its own fluctuations. Although their calculations yielded higher noise when side-modes were larger, the physical origin of the fitting term was not justified. This model was extended by Becher *et al.* [15] to include nonlinear gain terms. The nonlinear gain could explain asymmetry in the side mode intensities and noise with respect to the main mode. However, this model could not clarify the physical origin of the self-saturation which resulted in the excess noise in their calculations.

We have developed a model to explain the physical origin of degradation of the mode correlation. We calculated the intensity noise of a multi-mode laser from this model, in which, for the first time, the nonlinear gain and the nonlinear loss were both included [142]. The physical mechanisms responsible for the nonlinear gain are the spectral hole burning and the gain modulation caused by beating of modes [3]. The nonlinear loss is introduced by saturable absorbers present in the cladding region of a laser.

We obtained two conclusions from our calculations performed with this model [142]. One conclusion is that the nonlinear gain does not enhance noise even if the laser is oscillating in multiple longitudinal modes. This result follows from the conservation of energy. When a laser is operating far above the threshold, electronic excitation created by the noiseless pump is either converted into the laser emission or lost in the internal absorption. The identical loss for all longitudinal modes does not deteriorate anticorrelation among longitudinal modes. Therefore, the mode-partition noise should cancel out in the total intensity noise. The total intensity noise in this case is independent of the relative intensities of the side modes. Then, the total intensity noise should only be limited by the quantum efficiency.

The other conclusion is that the nonlinear loss does enhance the total intensity noise when side-mode intensities are higher. The power in the main mode saturates the nonlinear losses most effectively for the main mode but less effectively for the side modes. This mode-dependent loss degrades the anticorrelation among the longitudinal modes. Consequently, the total intensity noise is enhanced when side-mode intensities are large. In the calculations, when we used saturable absorption loss values which were estimated from known parameters of quantum-well (QW) lasers, the calculated noise quantitatively agreed with the experimental observations.

We computed the intensity noise of a multi-mode semiconductor laser using the Langevin rate equations. In this chapter we describe physical models for the nonlinear terms for the gain and the loss used in the rate equations. The method of solution of the rate equations is briefly reviewed. An intuitive explanation for the effect of the nonlinear loss on longitudinal-mode anticorrelation is also described.

To verify our model, we measured the wavelength-tuning characteristics of several lasers because their hysteresis behavior also depends on the amount of saturable absorption loss. We observed large hysteresis in the wavelength tuning of QW lasers. This behavior is strong evidence for the large amount of saturable absorption loss present in QW lasers. In comparison, no hysteresis was observed in transverse-junction-stripe (TJS) lasers. These observations were supported by the estimate of saturable loss from the doping concentration and profile and other parameters of the two kinds of lasers.

5.2 Physical Models for Nonlinearity

In the rate equation model, gain and loss are included as summed terms. In particular, we have incorporated nonlinear contributions to both the gain and the loss, which depend on the laser intensity or photon density. The linear terms are adopted from previous models.

5.2.1 Nonlinear Gain

The linear gain is modeled as a linear function of the carrier number and a quadratic function of the wavelength, i.e.,

$$G_{\mathrm{L},i} = A_i(N_{\mathrm{c}} - N_0) = A_0 \left(1 - \frac{\lambda_i}{\lambda_0}\right)^2 (N_{\mathrm{c}} - N_0) \quad , \tag{5.1}$$

where subscript i and 0 refer to the ith longitudinal side mode and the main mode, respectively, N_c is carrier number (note that it is not the carrier density), N_0 is the transparency carrier number, λ is the wavelength, and $A_i = \beta_i/\tau_{\rm sp}$ is the gain coefficient where β_i is the spontaneous-emission coupling factor and $\tau_{\rm sp}$ is the spontaneous emission lifetime. We neglect asymmetry in the gain with respect to the peak wavelength. An asymmetric gain profile would only lead to a quantitative change in the individual mode noise without affecting the fundamental qualitative behavior of the mode correlation and noise. For the nonlinear gain we adopt the following analytical expression derived by Agrawal $et\ al.$ [3]:

$$G_{\rm nl}(\omega_j) = -G_{\rm L}(\omega_0) \sum_k \zeta_{jk} n_k \quad . \tag{5.2}$$

Here the superscripts j, k refer to the longitudinal modes, ω is the angular frequency of a mode, and n the photon number (note that it is not the photon density). The gain saturation coefficients ζ_{jk} are given by the following expression

$$\zeta_{jk} = \frac{\mu^2 \omega_0 \tau_{\rm in} (\tau_c + \tau_v)}{V_{\rm m} 2\epsilon_0 \hbar \kappa \kappa_{\rm g}} \frac{G_{\rm L}(\omega_k)}{G_{\rm L}(\omega_0)} \frac{C_{jk}}{(1 + \delta_{jk}) C_k}$$
$$\times \left(1 + \frac{1 + \alpha_k \Omega_{jk} \tau_c}{1 + (\Omega_{jk} \tau_c)^2} \right) \quad , \tag{5.3}$$

where $\Omega_{jk} = \omega_j - \omega_k$ and α_k is a dimensionless parameter related to the slope of the gain curve at ω_k. The rate $1/\tau_{\rm in} = 1/\tau_c + 1/\tau_v$ contains the conduction band (τ_c) and the valance band (τ_v) relaxation times. C_{jk}/C_k is a factor arising from the spatial structure of the optical mode, $\delta_{jk} = 1$ if $j = k$ and zero otherwise, μ is the dipole moment matrix element, κ is the refractive index, $\kappa_{\rm g}$ is the group refractive index, and $V_{\rm m}$ is the optical mode volume. For a GaAs laser, $\mu = 4.6 \times 10^{-29}$ cm. C_{jk}/C_k and α_k are approximately equal to 1 and 0, respectively. The coefficients ζ_{ii} define the self-saturation and $\zeta_{ij}, i \neq j$, define the cross-saturation of the gain by the photon field. Substitution of the characteristic parameter values $\tau_{\rm in} = 0.1$ ps, $\tau_v = 0.07$ ps, $\tau_c = 0.2$ ps, $\kappa = 3.4$ and $\kappa_{\rm g} = 4$ in (5.3) results in $\zeta_{00} \times V_{\rm m} = 2.7 \times 10^{-18}$ cm^3. This value is in reasonable agreement with the value inferred from measurements of the self-saturation coefficient [3].

The self- and the cross-gain-saturation coefficients indicate that the gain at a lasing mode frequency is slightly reduced because of the finite intraband relaxation time of carriers. The amount of the reduction in the gain depends not only on the power of that mode (self-saturation), but also on the power of neighboring modes (cross-saturation). Beating between different longitudinal modes causes oscillation in the carrier density. This results in a small asymmetric contribution to the nonlinear gain. Note that in the context of semiclassical laser theory, the gain suppression is referred to as spectral hole burning, and dynamic variation in the carrier density is referred to as population pulsation.

5.2.2 Nonlinear Loss

Linear loss comes from free carrier absorption and scattering inside the cavity as well as leakage of power out of the cavity at the end mirrors. Then, the total linear loss coefficient is $\alpha_p = \alpha_{int} + \alpha_e$, where α_{int} is the internal loss and α_e is the external coupling loss coefficient. $\alpha_e = \frac{1}{L}\ln(1/R_1 R_2)$, where L is the length of the cavity and R_1, R_2 are facet reflectivities. In addition, saturable absorption caused by the deep trap levels in the band gap has been observed in AlGaAs materials [139, 171]. Deep traps, also known as DX centers, are formed by the lattice defect created by impurity atoms. Commonly used donor impurities such as Te [171] and Si [175] create DX centers in $Al_x Ga_{1-x} As$, which usually is the material of choice for the cladding layers in GaAs lasers. The saturable absorption coefficient by the DX centers, α_{sa}, derived from rate equations reads [46]

$$\alpha_{sa}(p) = \frac{\alpha_s}{1 + (p/p_s)} \quad , \tag{5.4}$$

with the unsaturated loss coefficient $\alpha_s = \sigma_o N_{DX}$ and the saturation photon density $p_s = N_c \sigma_c v_{th}/V \sigma_o v_g$. p is the photon density, σ_o is the optical absorption cross-section of the trap, σ_c is the cross section of capturing electrons from the conduction band into the trap, N_{DX} is the density of the DX centers, N_c/V is the density of the conduction-band electrons, v_{th} is the thermal velocity of the carriers in the conduction band, and v_g is the group velocity of the light in the material.

To obtain the saturable loss for a longitudinal mode, we assume that the saturation of traps comes mainly from one dominant mode. This is a good assumption when the side modes are a few times smaller than the main mode and the power in the main mode itself is a few times larger than the saturation intensity. The standing wave optical field of the main mode results in spatial variation of the saturation of traps inside the cavity. At the positions where the main-mode intensity is minimum, absorption by the traps is locally large. Since different longitudinal modes have a slightly different spatial profile, the average saturable loss for a side mode is slightly larger than the saturable loss of the main mode. The average saturable loss in a mode can be estimated by averaging the local loss weighted by the spatial intensity distribution of the mode [46]. When the main-mode intensity is about 2 to 10 times the saturation intensity, the saturable absorption loss of the nearest side mode could easily be about 10–20% larger than the main-mode loss.

The experimental value of σ_o for Si doping in $Al_{0.4} Ga_{0.6} As$ is $\sigma_o \approx 10^{-18}$ cm^2 and has negligible dependence on temperature. On the other hand, σ_c for Si in $Al_{0.4} Ga_{0.6} As$ strongly depends on temperature. The measured values are $\sigma_c \approx 10^{-26}$ cm^2 and $\sigma_c \approx 10^{-17}$ cm^2 at 80 K and 300 K, respectively. For the carrier density $N_c/V = 10^{17}$ cm^{-3}, the saturation photon density is $p_s \approx 7 \times 10^4$ photons/cm^3 and $p_s \approx 7 \times 10^{14}$ photons/cm^3 at 80 K and 300 K, respectively.

5.3 Noise Analysis Using Langevin Rate Equations

In the following we briefly describe the AC analysis of rate equations for three longitudinal modes, one main mode and two side modes symmetrically placed about the main mode. Using (5.2), (5.3) and the parameters discussed in the previous section, we rewrite the nonlinear gain as follows:

$$G_{\mathrm{nl},i} = -A_0 N_c \sum_k \zeta_{ik} n_k \tag{5.5}$$

$$\zeta_{ik} \approx -\frac{A_k}{A_0} \zeta_{00} \left\{ 1 + \frac{1}{1 + (\Omega_{ik} \tau_c)^2} \right\} . \tag{5.6}$$

We also define the linear gain of the side modes as $A_{1,-1} = \beta_{1,-1}/\tau_{\mathrm{sp}} = m\beta_0/\tau_{\mathrm{sp}} = mA_0$.

We assume that the nonlinear loss is saturated only by the main mode. We use a spatially averaged value of the saturable loss in Langevin rate equations and define the averaged expressions as

$$\alpha_{\mathrm{sa},i}^{av} = \frac{\alpha_{\mathrm{s},i}^{av}}{1 + (n_0/n_{\mathrm{s}})} \tag{5.7}$$

and

$$\alpha_{\mathrm{s},\pm1}^{av} = \alpha_{\mathrm{s},0}^{av}(1 + \Delta\alpha) \quad , \tag{5.8}$$

where $\Delta\alpha$ is the relative difference in the loss of the side mode and the main mode. The effective loss rate due to the saturable absorber then reads

$$\frac{1}{\tau_{\mathrm{sa},i}} = \alpha_{\mathrm{sa},i}^{av} v_g = \frac{1}{\tau_p} \frac{\alpha_{\mathrm{s},i}^{av}/\alpha_p}{1 + (n_0/n_{\mathrm{s}})}$$

$$= \frac{1}{\tau_p} \frac{\alpha_{\mathrm{r},i}}{1 + (n_0/n_{\mathrm{s}})} \quad , \tag{5.9}$$

where $1/\tau_p = \alpha_p v_g$ is the total linear loss rate. Then, $1/\tau_p = (\alpha_{\mathrm{int}} + \alpha_e)v_g = 1/\tau_{\mathrm{int}} + 1/\tau_e$. On this basis, the Langevin rate equations for the carrier number N_c and the photon number n_i in the ith mode ($i = 0, \pm 1$ for the main and side modes, respectively) are given by

$$\frac{dN_c}{dt} = P - \frac{N_c}{\tau_{\mathrm{sp}}} - \sum_i A_i N_c n_i + A_0 N_c \sum_k \zeta_{ik} n_k n_i$$

$$+ \Gamma^p + \Gamma^{sp} + \Gamma^{st} \tag{5.10}$$

$$\frac{dn_i}{dt} = -\frac{n_i}{\tau_p} - \frac{n_i}{\tau_p} \frac{\alpha_{\mathrm{r},i}}{1 + n_0/n_{\mathrm{s}}} + A_i N_c(n_i + 1) - A_0 N_c \sum_k \zeta_{ik} n_k n_i$$

$$+ f_i^g + f_i^l + f_i^e, \tag{5.11}$$

where P is the pump rate and Γ^p, Γ^{sp} and Γ^{st} are the Langevin noise sources for the carrier number due to fluctuations in the pump, the spontaneous emission and the stimulated emission, respectively. Similarly f_i^g, f_i^l and f_i^e are Langevin noise sources for the photon number due to fluctuations in the gain, the internal loss and the external coupling loss, respectively.

We did not solve the DX-center rate equation simultaneously. The DX-center dynamics can be adiabatically eliminated because the dynamics of deep traps ($\sim$10 ns) is 3 to 4 orders of magnitude slower than the photon lifetime ($\sim$1 ps). The rate equation for the deep traps was solved separately in [46], and the result is used here to eliminate the trap density variable in the saturable loss term in the photon rate equation.

The correlation functions for all these six noise sources, as derived in [105], are given by

$$\langle \Gamma^p(t)\Gamma^p(t')\rangle = 0 \tag{5.12}$$

$$\langle \Gamma^{sp}(t)\Gamma^{sp}(t')\rangle = \delta(t-t')\langle N_c\rangle/\tau_{sp} \tag{5.13}$$

$$\langle \Gamma^{st}(t)\Gamma^{st}(t')\rangle = \delta(t-t')\sum_i(\langle G_i^l\rangle + \langle G_i^{nl}\rangle)\langle n_i\rangle \tag{5.14}$$

$$\langle f_i^g(t)f_i^g(t')\rangle = \delta(t-t')(\langle G_i^l\rangle + \langle G_i^{nl}\rangle)\langle n_i\rangle \tag{5.15}$$

$$\langle f_i^l(t)f_i^l(t')\rangle = \delta(t-t')\langle n_i\rangle/\tau_{lo,i} \tag{5.16}$$

$$\langle f_i^e(t)f_i^e(t')\rangle = \delta(t-t')\langle n_i\rangle/\tau_e \tag{5.17}$$

$$\langle \Gamma^{st}(t)f_i^g(t')\rangle = -\delta(t-t')(\langle G_i^l\rangle + \langle G_i^{nl}\rangle)\langle n_i\rangle \tag{5.18}$$

where $\langle\ \rangle$ denotes ensemble average, and $1/\tau_{lo,i} = (\alpha_{int} + \alpha_{sa,i})v_g = 1/\tau_{int} + 1/\tau_{sa,i}$ is the total internal loss rate. The remaining combinations of noise sources that have not been included in the above list are uncorrelated. The pump fluctuation, $\Gamma^p(t)$ is taken to be zero because of the assumption of a noiseless pump source. Since both $\Gamma^{st}(t)$ and $f_i^g(t)$ originate from the same source – fluctuations in the stimulated emission in which the loss of a carrier results in the emission of a photon and vice versa – these two noise sources are perfectly anticorrelated, as in (5.18).

Steady-state values of the photon number in all the three modes, $\langle n_i\rangle$, were calculated from (5.10) and (5.11) by substituting zero for the time-derivative term and noise terms. The photon-number noise was derived by performing small-signal AC analysis of the rate equations, employing the method used in Chap. 2. To summarize, variable decompositions, $N_c = \langle N_c\rangle + \Delta N_c$ and $n_i = \langle n_i\rangle + \Delta n_i$, were substituted in the rate equations. Then, the resulting equations were linearized in fluctuating components. Fourier transforms of these equations were solved to find the expressions for the fluctuating variables as functions of the noise sources. Fourier-analysis frequency was taken to be zero in our calculations. This derivation gave fluctuations in the photon number inside the cavity, Δn_i, which were then expressed in terms of the photon field amplitude, a_i ($n_i = a_i^2$), as $n_i = \langle n_i\rangle + \Delta n_i = \langle a_i\rangle^2 + 2\langle a_i\rangle\Delta a_i$ or $\Delta n_i = 2\langle a_i\rangle\Delta a_i$. The noise in the photon flux outside the cavity, Δr_i, can be expressed in terms of Δa_i using the boundary condition at the output coupling mirror [161],

$$\Delta r_i = \frac{1}{\sqrt{\tau_e}} (\Delta a_i) - \sqrt{\tau_e} \frac{f_i^e}{2 \langle a_i \rangle}. \qquad (5.19)$$

The second term in the above equation is the vacuum field reflected from the cavity mirrors. We then combined the noise of all the three modes in the output flux and calculated the total intensity noise. This corresponds to the noise which is measured in experiments.

5.4 Numerical Results

5.4.1 Numerical Parameters

The parameters used in calculations were $\tau_{sp} = 1$ ns, $\tau_p = 1$ ps, $\tau_{int} = 1.33$ ps and $\tau_e = 4$ ps. The pumping level was 10 times the threshold. The relative gain parameter, $m = \beta_i / \beta_0$, was varied in the range 0.99–0.99999 to change the ratio of the power in the side modes with the power in the main mode. In the following we will refer to this ratio as the side-mode-suppression ratio (SMSR). The value of β_0 was either fixed as 2.6×10^{-6} or varied in the range $10^{-6} - 10^{-4}$ to change the mode-partition noise for the same SMSR. For the self-saturation coefficients, we used $\zeta_{00} \times V_m = 2.7 \times 10^{-18}$ cm^3 [3] and deduced the remaining nonlinear gain coefficients from (5.6). The optical mode volume V_m was 6.2×10^{-10} cm^3. The average unsaturated nonlinear loss coefficient α_s^{av} was approximated as $\alpha_s^{av} \approx \sigma_o N_{DX}$. The optical absorption cross-section of a DX center (σ_o) was taken from measurements on Al$_x$Ga$_{1-x}$As doped with Si [175], which is commonly used as donor impurity in Al$_x$Ga$_{1-x}$As lasers. The measured value was $\sigma_o = 10^{-18}$ cm^2 for 0.9 μm laser light. This value was found to be insensitive to the Si concentration and alloy composition change. It has also been observed that the concentration of DX centers (N_{DX}) is almost the same as the concentration of the Si dopant. To see the influence of the doping concentration on the intensity noise, we varied the value of N_{DX} in the range $10^{17} - 10^{18}$ cm^{-3}. For $N_{DX} = 10^{17}$ cm^{-3}, if we assume $N_c / V = N_{DX}$, the saturation photon density $p_s \approx 10^{16}$ mW/cm^3 at 300 K (for 0.9 μm light). Then, the average loss for the main mode is, $\alpha_{sa,0}^{av} \approx 0.01$ cm^{-1}. We assumed the relative absorption parameter $\Delta\alpha = 0.09$, which is close to the estimate obtained from physically reasonable parameters [139].

5.4.2 Results

To elucidate the influence of the nonlinear gain and saturable loss on the intensity noise, we started our analysis by calculating noise in the absence of both the nonlinear gain and the saturable loss. Figure 5.1 displays the

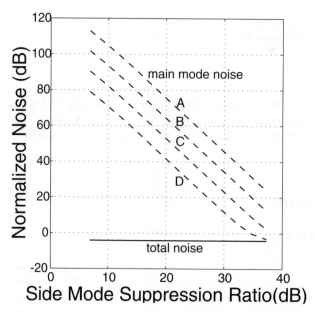

Fig. 5.1. Intensity noise versus SMSR for the case with only linear gain and linear loss. Noise was normalized by the shot noise. The spontaneous-emission coupling factor, β_0, was varied as follows: (A) 2×10^{-6}, (B) 7.36×10^{-6}, (C) 2.7×10^{-5}, and (D) 1×10^{-4}; other parameters were fixed as follows: the pumping level $I/I_{th} = 10$, the external quantum efficiency $\eta_e = 0.75$, and the total linear loss $\alpha_p = 120\,\mathrm{cm}^{-1}$

result, and, as expected, the total intensity noise did not depend on the size of the side modes (SMSR); however, the noise of an individual (main) mode strongly depended on the SMSR. The total intensity noise was determined by the internal loss (or the quantum efficiency, QE) and the pumping level. This value is referred to as the QE limit of the intensity noise squeezing. The noise of the main mode was about 80 dB larger than the total intensity noise when SMSR was about 20 dB.

In the next step, we repeated the calculations with only the nonlinear gain (Fig. 5.2) term. The total intensity noise remained insensitive to the SMSR. The total intensity noise and the main mode noise were the same as in the linear case (Fig. 5.1).

Figure 5.3 shows the noise when only saturable absorption loss was included. In this case, the excess total intensity noise was present when the SMSR was small. The total intensity noise decreased and approached the QE limit as the SMSR was increased. It is evident from this analysis that the saturable absorption enhances the total intensity noise when the mode-partition noise is large. Finally, when we included both the nonlinear gain and the saturable loss in our calculations, the total intensity noise had the same dependence on the SMSR as obtained with the saturable loss alone (Fig. 5.4).

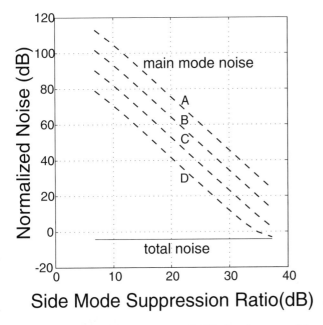

Fig. 5.2. Intensity noise versus SMSR for the case with nonlinear gain and linear loss. The self-gain-saturation coefficient was $\zeta_{00} = (1/V_m) \times 2.7 \times 10^{-18}$ (V_m is the optical mode volume in cm^3) and other parameters were the same as in Fig. 5.1

These results are in contrast with the common hypothesis that the gain nonlinearity increases the total intensity noise of a multi-mode laser [15]. Our calculations show that nonlinear gain does not enhance the total intensity noise of the laser. It is the nonlinear loss which introduces excess noise into a multi-mode laser.

5.5 Discussion: Effect of Saturable Loss

The results presented in Sect. 5.4 can be intuitively interpreted as follows: The nonlinear gain saturates the conversion of electronic excitation supplied via a quiet pump to the cavity-mode photons. However, when the dominant mode is far above the threshold, photons either go out of the cavity in any of the cavity modes or are deleted equally from all the modes because of the linear loss. Since it is the same gain which is shared by all the longitudinal modes, the mode-partition noise of each mode should still be perfectly anticorrelated with rest of the modes, collectively. Hence, the mode-partition noise does not contribute to the noise of the total photons generated inside the cavity. Identical loss for the modes does not affect this null contribution. Therefore, the total laser noise detected outside the cavity should approach

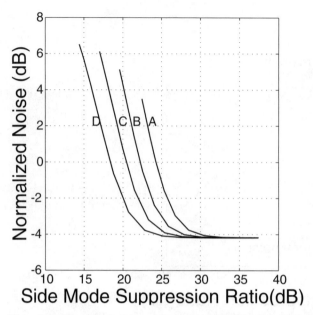

Fig. 5.3. Total intensity noise versus SMSR for the case with linear gain and nonlinear loss. The saturable loss coefficient $\alpha_{\text{sa},0}^{av}$ was varied as follows: (A) $10^{-3} \times \alpha_p$, (B) $0.5 \times 10^{-3} \times \alpha_p$, (C) $0.25 \times 10^{-3} \times \alpha_p$, and (D) $0.125 \times 10^{-3} \times \alpha_p$, where $\alpha_p = 120\,\text{cm}^{-1}$ was the total linear loss. The relative excess saturable loss for the side modes was $\Delta\alpha = 0.09$, the spontaneous-emission coupling factor was $\beta_0 = 2.6 \times 10^{-6}$, and the rest of the parameters were the same as in Fig. 5.1

the QE limit, independent of the SMSR. Alternatively, for an intuitive understanding, the fluctuations in the photon number of the ith mode inside the cavity, Δn_i, can be loosely defined as $\Delta n_i = \Delta n_i^p + \Delta n_i^m$, where the contribution Δn_i^p is determined by the pumping level and linear loss, and Δn_i^m is determined by the mode competition or the mode-partition noise. In this context, anticorrelation means that $\sum_i \Delta n_i^m = 0$. The noise observed outside the cavity with the external QE, η_e, is thus $\eta_e \sum_i \Delta n_i = \eta_e \sum_i \Delta n_i^p$. This result is the same as that of a single-mode laser for the same pumping level.

However, the saturable absorber makes the loss of each longitudinal mode dependent on the power of the dominant mode. When the main mode fluctuates to a higher power level, the gain is reduced, and, consequently, the emission in all other modes decreases. This is the origin of the intensity noise anticorrelation among the longitudinal modes. The fluctuations of the main mode also reduce the saturable loss of the side modes. The reduction in the internal loss in turn increases the output coupling efficiency of power in the side modes. This process partly compensates for the decrease caused by the smaller gain. Since the nonlinear loss for the main-mode is heavily saturated by its large intensity, the fluctuation in the main mode power has a negligible effect on its output coupling efficiency. Thus, the fluctuation in

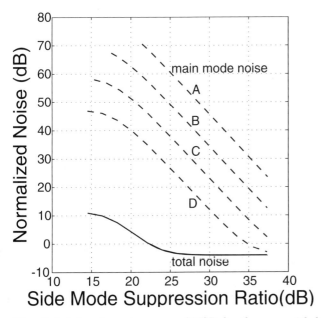

Fig. 5.4. Intensity noise versus SMSR for the case with both nonlinear gain and nonlinear loss included. The spontaneous emission coupling factor, β_0, was varied as follows: (A) 2×10^{-6}, (B) 7.36×10^{-6}, (C) 2.7×10^{-5}, and (D) 1×10^{-4}. The self gain saturation coefficient was $\zeta_{00} = (1/V_m) \times 2.7 \times 10^{-18}$ (V_m is optical mode volume in cm^3), the saturable loss coefficient was $\alpha_{sa,0}^{av} = 0.5 \times 10^{-3} \times \alpha_p$, the relative excess saturable loss for the side modes was $\Delta\alpha = 0.09$, and the rest of the parameters were the same as in Fig. 5.1

the main-mode power modulates the output coupling efficiency for the side modes but not for the main mode. Consequently, outside the laser cavity the overall anticorrelation between the main mode and the side modes becomes less perfect. Therefore, the mode-partition noise does not cancel out when all the modes are detected.

When the laser is operating far above the threshold and the pump is noiseless, the total photon flux generated internally should be noiseless, because the carrier fluctuations cancel out due to the conservative coupling between photons and carriers. However, some photons get lost internally (by absorption and scattering), and only those photons which escape the laser cavity are detected. One might expect that these events are random and therefore contribute Poissonian noise to the output photon flux measured in the experiments. Thus the total intensity noise would only be limited by the QE. This assumption is no longer valid for any loss which depends on the intensity of the photon flux. In this case loss of photons is no longer a truly independent event. Therefore, the fluctuations in the output photon flux and the photons which are lost (we call it lost photon flux) are correlated. This correlation changes the intrinsic anticorrelation among longitudinal modes (originating from the shared gain) in the output photon flux and results in excess noise when the mode-partition noise is large. In the following is a simple mathematical description of this argument. If we assume all modes to have identical saturable loss ($\Delta\alpha = 0$), the average rate of photon loss (lost photon flux), Φ_l, may be expressed as

$$
\begin{aligned}
\Phi_l &= \sum_i \left(\frac{1}{\tau_{\text{int}}} + \frac{1}{\tau_{\text{sa}}} \right) n_i \\
&= \sum_i \left(\frac{1}{\tau_{\text{int}}} + \frac{\alpha_r}{\tau_p[1 + (n_0/n_s)]} \right) n_i \quad .
\end{aligned}
\tag{5.20}
$$

Then, the fluctuations in the lost photon flux, $\Delta\Phi_l$, read

$$
\begin{aligned}
\Delta\Phi_l &= \sum_i \left(\frac{1}{\tau_{\text{int}}} + \frac{\alpha_r}{\tau_p[1 + (n_0/n_s)]} \right) \Delta n_i \\
&\quad - \sum_i \frac{\alpha_r}{\tau_p[1 + (n_0/n_s)]^2} \frac{\Delta n_0}{n_s} n_i \\
&= \left(\frac{1}{\tau_{\text{int}}} + \frac{\alpha_r}{[1 + (n_0/n_s)]} \right) \sum_i \Delta n_i \\
&\quad - \frac{1}{\tau_p} \frac{\alpha_r}{[1 + (n_0/n_s)]^2} \frac{\Delta n_0}{n_s} \sum_i n_i \quad .
\end{aligned}
\tag{5.21}
$$

The first term in the above expression is proportional to the noise of the total internally generated photons $\sum_i \Delta n_i$, which depends only on the pumping level. It should be insensitive to the SMSR because the mode-partition noise in the each mode is canceled by the anticorrelated noise in the other modes. The second term is proportional to the internal noise in the main mode, which is large when the side-mode intensities are bigger. The negative sign of this term should be noted. It implies that the noise of the lost photon flux is anticorrelated with the main-mode noise.

This simple description was verified by more exact calculations using (5.10) and (5.11). Figure 5.5 shows that the noise of the lost photon flux is large when the SMSR is large. The noise of the combined flux, including the output flux (the total external intensity noise) and the lost photon flux, is also plotted in the Fig. 5.5. The noise of the combined flux is smaller than the noise of both the lost photon flux and the output photon flux, which indicates that the noise of the lost photon flux is anticorrelated with the noise of output photon flux. Again, we emphasize that it is the total external intensity noise which is measured in an experiment. The observed noise is larger than the QE limit because of the deletion of the anticorrelated noise which is carried by the lost photon flux. The noise of the combined flux or internal photon flux is independent of the SMSR and is determined by the pumping level only.

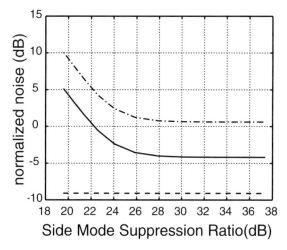

Fig. 5.5. Intensity noise of the lost photon flux (*dot-dashed line*), the output flux (*solid line*), and the combined flux (*dashed line*) versus SMSR

These calculation results imply that the so-called excess noise (noise higher than the QE limit) measured in many lasers stems from the non-linear loss processes. The amount of excess noise depends on the strength of the nonlinear loss and the intensity of the side modes.

5.6 Comparison of Two Laser Structures with Respect to Saturable Loss

The numerical results can explain some of the differences in the noise characteristics of different types of lasers. Specifically, we consider the difference in the behavior of TJS and QW edge emitting lasers. As we mentioned in Sect. 5.2, the donor impurities in the cladding layer of AlGaAs lasers create DX centers which behave like a saturable absorber. We estimated the saturable absorption loss in the two lasers from the known concentration of the silicon doping and the free carriers [143].

5.6.1 Estimate of the Loss by Si DX Centers

The first important difference between the two types of lasers is in the current-injection scheme. In a TJS laser, the active junction is a homojunction located within an active layer perpendicular (transverse) to the growth direction. The pump current is injected in the active layer plane, as shown in Fig. 5.6a. The doping concentration of the cladding layers in a TJS laser is typically $\sim 10^{17}\,\mathrm{cm}^{-3}$.

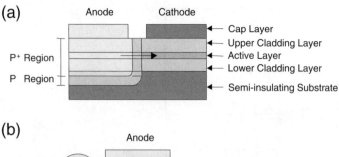

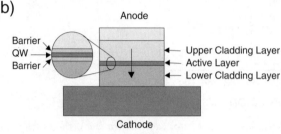

Fig. 5.6 a, b. Schematic structure and the current-injection directions for (**a**) a TJS laser and (**b**) a QW laser. The *arrows* indicate the direction of current flow

In contrast, the p–n junction in a QW laser is formed along the growth direction, by successively growing the p-type and n-type layers. In a popular double-heterostructure QW laser, separate layers of AlGaAs with different Al concentrations are used as barriers (to confine carriers) and cladding layers (to confine the optical field). The current flows perpendicular to the growth planes (Fig. 5.6b), and the cladding layers and the barrier layers need to be heavily doped ($10^{18} \sim 10^{19}$ cm^{-3}) to reduce the series resistance. Due to the higher doping levels of the cladding layer, the DX-center concentration in a QW laser is higher than that in a TJS laser. The saturable absorption by a DX center increases monotonically with the concentration of DX centers (N_{DX}).

The saturation photon density increases with the concentration of free carriers which can be trapped by these centers. In a QW laser the free carrier concentration in the cladding layer, which injects carriers into the active junction, is similar to the doping concentration. On the other hand, the part of the cladding layer of a TJS laser where the optical field spreads is also a transverse p–n junction. Since free carriers are depleted from this region, the concentration of the conduction-band electrons is significantly lower than the doping concentration, which is already an order of magnitude smaller than that of a QW laser. Consequently, the strength of saturable absorption in a TJS laser should be more than two orders of magnitude smaller than in a QW laser.

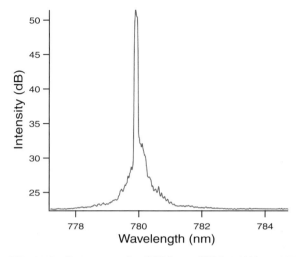

Fig. 5.7. Spectrum of a QW laser (SDL 5400) at 290 K and 86 mA bias current (threshold = 28 mA)

5.6.2 Experimental Verification of the Saturable Loss

The saturable loss manifests itself in two distinct spectral characteristics of lasers [143]. First, the side-mode suppression becomes higher because the weaker modes suffer a larger loss than the stronger mode. Second, the change in wavelength with the change in temperature or pump current exhibits larger hysteresis. These two properties can be measured experimentally to compare the level of saturable loss in different types of lasers.

First, we compare the optical spectra of QW and TJS lasers. The intensities of the side modes in QW lasers are typically about 20 dB below that of the main mode (Fig. 5.7), while the side-mode intensities in TJS lasers are only 2 to 5 dB below the main-mode intensity (Fig. 5.8). This is an indication that the saturable loss is larger in QW lasers.

In a second experiment, we measured the change in the main-mode wavelength for several lasers as the pump current was ramped up and back down. The measurements were done at room temperature [143]. Figure 5.9 shows the measurement results for a single QW laser (Spectra Diode Lasers, 5410 series) with a threshold current of 28 mA. All the QW lasers we measured exhibited similarly large hysteresis in the tuning characteristics. As the pump current was increased, the dominant mode, instead of moving to the adjacent mode, hopped a few modes farther away from it. The change in the mode number with the mode hopping was even larger when the pump current was decreased. The area of the hysteresis loop was larger at higher pump currents. In some lasers, under similar operating conditions, the main mode hopped to as far as five modes away. The hysteresis behavior can be attributed to the large amount of saturable absorption loss. This is because the saturable

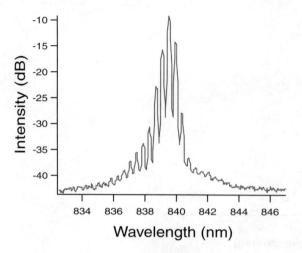

Fig. 5.8. Spectrum of a TJS laser at 80 K and 31.1 mA bias current (threshold = 1.5 mA)

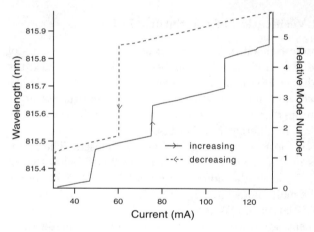

Fig. 5.9. The wavelength of the main mode versus current of a QW laser (SDL 5410) with a threshold current of 28 mA and a longitudinal-mode separation of 0.11 nm at 290 K. The right vertical axis is the relative change in the mode number with respect to the position of the main mode at the threshold (indicated by 0)

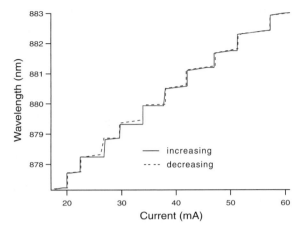

Fig. 5.10. The wavelength of the main mode versus current of a TJS laser (L475) with a threshold current of 15.8 mA and a longitudinal-mode separation of 0.48 nm at 290 K

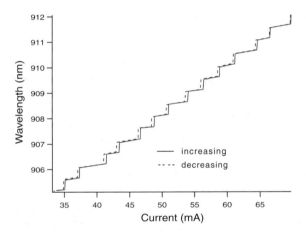

Fig. 5.11. The wavelength of the main mode versus current of a laser TJS (L78) with a threshold current of 26.8 mA and a longitudinal-mode separation of 0.48 nm at 290 K. This laser had larger concentration of Si compared to the TJS laser (L475) shown in Fig. 5.10

loss tries to stabilize the mode which is already dominant (by reducing its loss) even when the gain peak shifts due to the change in current and/or temperature.

Figure 5.10 shows similar measurement results for a TJS laser (L475) with a threshold current of 15.8 mA. As shown, this laser did not show any hysteresis characteristics. Figure 5.11 shows the measurement results for a L78 series TJS laser, which has a higher concentration of silicon impurity but the same carrier concentration as that of L475 lasers. This laser exhibited a little hysteresis. Both L475 and L78 series TJS lasers were grown at Hamamatsu Photonics Central Research Lab. Details about the growth and noise characteristics of L475 and L78 lasers are presented in Chap. 6. L78 lasers did not show squeezing, while L475 lasers exhibited almost QE-limited squeezing.

The observation of the large hysteresis in QW lasers and the negligible hysteresis in TJS lasers is strong evidence for the existence of the larger saturable absorption loss in the former. The difference in the two TJS lasers with different amount of silicon doping also supports the fact that saturable loss is reduced by decreasing the Si concentration. As noted earlier, the L78 series TJS lasers (with higher Si concentration) had higher noise compared to the L475 series TJS lasers (with lower Si concentration). The lower noise with the smaller saturable loss clearly indicates that the saturable absorption loss created by Si dopant enhances the intensity noise.

5.6.3 Explanation for the Excess Noise in QW Lasers

The higher concentration of both Si and free carriers in the AlGaAs cladding layers of QW lasers indicates the larger saturable absorption loss in these layers compared to the TJS lasers. This was verified by the observation of hysteresis behavior in the wavelength-tuning characteristics of the QW lasers. We then checked the quantitative influence of this loss on the noise. In fact, the calculations described in Sect. 5.5 employed the same value of saturable loss that was estimated from Si and the free-carrier concentration in QW lasers. Indeed, as shown in Fig. 5.3, the noise was higher for smaller SMSR. Furthermore, a value of a few dB of excess noise around 20 dB SMSR and the QE-limited squeezing around 35 dB SMSR seen in this figure agree well with the experimental observations summarized in Table 4.1 of Chap.4.

When we used the value of the saturable loss estimated from the doping of TJS lasers, calculated noise was independent of the SMSR and virtually identical to that expected from the QE limit. This implied that in the case of TJS lasers the saturable loss is too small to significantly influence the noise.

The agreement between the experiments and the calculations based on our model strongly suggest that it is the saturable loss which is responsible for the excess noise and the degraded mode correlation observed in QW lasers.

5.7 Summary

We calculated the intensity noise on the basis of Langevin rate equations for a semiconductor laser with three longitudinal modes. In our model, we have incorporated both nonlinear gain and nonlinear loss terms. The mechanisms considered for the nonlinear gain are spectral hole burning and population pulsations. Saturable absorption by the DX center is included as the nonlinear loss mechanism.

Our calculations show that for a multi-mode laser the nonlinear gain alone does not lead to enhanced total intensity noise in comparison with the linear model. This is because the nonlinear gain preserves the anticorrelation among the longitudinal modes. Hence, the mode-partition noise cancels out in the total intensity noise.

It is the nonlinear loss which leads to the excess noise observed in a multi-mode laser. In the case of saturable loss, the fluctuations in the laser output are anticorrelated with those of absorbed photons. The removal of this anticorrelated noise increases the noise measured at the laser output.

We have found strong experimental evidence and theoretical support for higher saturable loss being the origin of the excess noise behavior observed in QW lasers. The cladding layers of QW lasers contain high concentrations of DX centers (created by Si dopant) and free carriers. We observed large hysteresis in the wavelength-tuning characteristics of QW lasers. This observation suggested a large amount of saturable loss in QW lasers. The noise calculated using the estimated value of the saturable loss present in typical QW lasers was in good agreement with the experimental observations. Unlike QW lasers, TJS lasers showed negligible hysteresis in the wavelength-tuning characteristics. The calculations also predicted no excess noise for TJS lasers.

In a practical device both kinds of nonlinearities exist. In Sect. 5.2 we described mechanisms of nonlinearities. We mentioned the absorption by DX centers as one of the mechanisms of the saturable loss. We clarify that our simulation model does not represent an accurate quantitative analysis of a practical multi-mode laser. However, in our opinion, it covers the basic qualitative features and thus helps in shedding light on the physical mechanisms of intensity noise generation. The results presented here should remain qualitatively valid when more rigorous models for the gain and the loss are incorporated into spatially resolved simulations.

6. Transverse-Junction-Stripe Lasers for Squeezed Light Generation

6.1 Introduction

The noise characteristics of semiconductor lasers as seen in experiments strongly depend on laser structure. In general, transverse-junction-stripe (TJS) lasers demonstrate lower noise and more often produce squeezing. The maximum squeezing was observed in the TJS lasers which were made from wafers grown by an old liquid-phase epitaxy (LPE) technique [198]. This technique has become obsolete, and wafers are now grown using the metal–organic chemical vapor deposition (MOCVD) technique. MOCVD-grown lasers showed higher noise than the LPE-grown lasers. Until recently, the MOCVD process had not been optimized for the generation of a high level of squeezing.

Quantum-well (QW) lasers, which became prevalent in later years, usually have noise higher than the shot-noise level, although careful experiments have realized amplitude squeezing in these lasers as well [120]. As described in Chap. 4 and 5, the excess noise originates from the residual longitudinal-mode-partition noise, more specifically, the mode-partition noise of a multi-mode laser which does not cancel out because of the imperfect anticorrelation among the longitudinal modes. In contrast to QW lasers, TJS lasers produce squeezing even when the spectrum is highly multi-mode [105]. This implies that TJS lasers have perfect intensity noise correlation among longitudinal modes.

Another suitable characteristic of TJS lasers in generating squeezing is their very low threshold at cryogenic temperatures, which allows operation far above the threshold of the laser. In this case the intensity noise due to spontaneous emission becomes insignificant, and squeezing should only be limited by the quantum efficiency (QE).

The low noise behavior and the small threshold current of TJS lasers motivated us to improve the MOCVD process used to grow lasers, in order to reproducibly achieve a large amount of squeezing . We modified the fabrication method to reduce defects in the material in order to decrease the internal loss [143]. We also tailored the doping profile of the laser active region to achieve good optical and electrical confinement. Both the smaller internal loss and the better optical confinement realized a low threshold current and high QE. These characteristics are desirable but not sufficient for squeezing. The important processing step which we implemented was use of a high As-to-Ga ratio (known as a V/III ratio) during MOCVD growth in order to reduce the Si doping concentration incorporated into the AlGaAs cladding layers. As described in Chap. 5, the Si impurity seems to be the main cause of noise enhancement in multi-mode lasers [142].

In this chapter the development of the fabrication method and the noise characterization of TJS lasers are discussed. These lasers consistently produced about −4 dB of squeezing [143]. The lasers were fabricated at Hamamatsu Photonics K. K., Central Research Laboratories, Hamamatsu, Japan, and were characterized at Stanford University.

6.2 Fabrication

We used the MOCVD method to grow the epitaxial layered structure of a TJS laser shown in Fig. 6.1. A lower cladding AlGaAs layer, an active GaAs layer, an upper cladding AlGaAs layer, and a GaAs cap layer were successively grown on a semi-insulating GaAs substrate. Thickness, carrier concentration and Al composition of these layers are given in Table 6.1. All these layers were made n-type by Si doping during growth. The upper cladding and the cap layer were heavily doped in order to reduce current leakage and bulk resistance.

6.2.1 Si Diffusion and Intermixing

The Si impurity in GaAs–AlGaAs hetrostructures is known to enhance diffusion of Al, which results in the intermixing of layers and lattice vacancies. The intermixing of the active and the cladding layers weakens the optical confinement provided by the difference in refractive index. The interface roughness

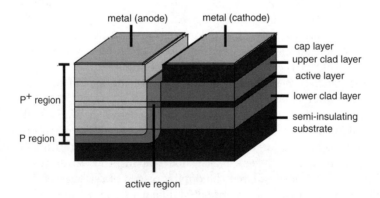

Fig. 6.1. Schematic of a TJS laser structure which was grown by MOCVD. The darker region is n-type

Table 6.1. Thickness, composition and carrier concentration in the different layers of the TJS laser shown in Fig.6.1

Layer	Composition	Thickness (μm)	Doping (cm^{-3})
Lower cladding	$Al_x Ga_{1-x} As$	1.5	10^{16}
Active	GaAs	0.075–0.1	2×10^{18}
Upper cladding	$Al_x Ga_{1-x} As$	1.5	10^{17}
Cap	GaAs	1.0	2×10^{18}

increases the scattering loss for photons. The Si impurity atoms also create deep traps for electrons (DX centers) in AlGaAs which act as saturable absorbers for photons. Consequently, the effective gain becomes smaller, and the internal loss becomes larger. More importantly, the intensity noise correlation between different longitudinal modes is degraded by the saturable absorption [142].

The mechanism for the enhancement of Al diffusion is as follows [170]: Two Si atoms residing on the nearest-neighbor donor–acceptor sites form a neutral Si pair which is assumed to be a mobile diffusion species [82]. Individual Si atoms are relatively immobile, since their motion involves charge exchange. The motion of the neutral Si pair creates di-vacancies which facilitates the diffusion of Al atoms. This leads to intermixing of layers. Because of the higher concentration of Si in the upper cladding layer, the intermixing is more prominent between this layer and the active layer.

6.2.2 High V/III Ratio for Sharper Interfaces

From the model of Si-assisted Al diffusion, we expected that the number of intermixing defects and deep traps would be reduced if we could decrease the Si concentration. The Si atoms substituting Ga(Al) at the lattice sites donate electrons to the crystal. Therefore, increased incorporation of Si atoms at the Ga(Al) sites would require a lower Si concentration to achieve the given carrier concentration. In order to do so, we used a high ratio of As to Ga/Al (V/III ratio) during growth. Then, it was possible to obtain the required carrier density with a lower Si concentration. The measurement of the emission spectrum confirmed that intermixing of the active and the cladding layers was significantly reduced.

6.2.3 P Doping by Zn Diffusion

After growing the n-type layered structure, p doping was obtained by diffusing Zn into the structure. Zn was diffused at 750°C to achieve heavy p doping (p$^+$) with a carrier density of about 10^{20}/cm^3. The wafer was then

Table 6.2. The DC characteristics of the TJS lasers fabricated with different V/III ratios. The front-facet reflectivity was $R_1 = 10\%$, and measurement temperature was 293 K.

Property	V/III ratio	
	78	156
Threshold current	29.6 mA	14.5 mA
Internal loss	27.6 cm^{-1}	12.2 cm^{-1}
Internal quantum efficiency	0.65	0.955
Differential efficiency	0.41	0.75
Corrected squeezing	$\geq$0 dB	-2.4 dB

heated to 900°C. The Zn atoms (and ions) close to the diffusion front, diffused further inside and formed an about 2 μm thick p region with a carrier density of $10^{19}/\text{cm}^3$. The change in the refractive index due to the change in the carrier density at the p^+–p boundary provided a weak lateral confinement of the optical field. Additionally, the higher concentration of impurity ions enhanced the carrier-scattering rate. Thus, the intra-band thermalization became faster, which resulted in more homogeneous gain broadening.

6.2.4 Devices

The grown and processed wafer was then cleaved into smaller chips and dielectric coating was deposited on both facets. The coating was antireflection (AR) and high reflection (HR) at the front and the rear facets, respectively. The length of the laser cavity and the front facet reflectivity were varied in order to optimize squeezing. Some of the laser chips were mounted on an SiC submount, which was put on a copper heat sink.

6.3 DC Characterization: Threshold, Loss and Quantum Efficiency

The threshold current for the lasers with 200 μm length and 10% AR coating was about 16 mA at room temperature. The leakage current through AlGaAs p–n junctions in the cladding layers which are parallel to the active p–n junction and nonradiative losses are significant at room temperature. Both the leakage current and the nonradiative losses decrease with decreasing temperature, and a lower threshold is realized at lower temperatures. At 80 K the threshold current was 1.6 mA and the external QE was about 0.75. This means that the threshold current density was 10 kA/cm^2, which is comparable to some of the best values obtained so far with the TJS laser structure.

The external differential QE (η_d) of a laser is given by

$$\frac{1}{\eta_d} = \left(1 + \frac{\alpha_i L}{\ln\left(\frac{1}{R_1 R_2}\right)} \right) \times \frac{1}{\eta_i} \quad , \tag{6.1}$$

where η_i is the internal quantum efficiency, α_i is the internal loss, L is the length of the cavity, and R_1 and R_2 are the reflectivities of the front and the rear facet, respectively. We measured η_d of lasers with different cavity lengths. We then extracted the internal loss (α_i) and the internal QE (η_i), from the intercept and the slope, respectively, of the linear fit of $1/\eta_d$ with L. The values of α_i and η_i corresponding to the V/III ratios of 78 and 156 are presented in Table 6.2. A higher V/III ratio resulted in lower α_i and higher η_i, clearly showing the reduction in the density of defects. Both were further improved in lasers grown with a V/III ratio = 475.

6.4 Intensity Noise

The intensity noise of these lasers was measured by the double balanced homodyne detection technique [163]. This technique allowed simultaneous measurement of the laser intensity noise and the corresponding shot noise. Large-area (1 cm^2) p–i–n Si photodiodes (Hamamatsu S3994) and low-noise radio-frequency (RF) amplifiers (Avantak UTC-517) were used in the detection setup. We measured noise at around 10 MHz using an RF spectrum analyzer. The measurement setup had an about 25% overall loss including a 95% QE of the detector. When we correct the data for this external loss, they corresponded to the squeezing value at the laser output facet.

6.4.1 Influence of High V/III Ratio

The noise of the first batch of lasers grown with a V/III ratio equal to 78 (sample name: L78) was either above or equal to the shot-noise level. The lasers grown with the higher V/III ratios, 156 (sample name: L156) and 475 (sample name: L475), could produce squeezing.

Table 6.3 compares the measurement temperature, threshold, measured external QE (η_e) and the best squeezing of the L156 and L475 lasers. Note that the lasers with the higher V/III (L475) ratio show smaller noise, or larger squeezing. L156 lasers showed a best squeezing of −1.5 dB at an optimal temperature of 140 K. When we corrected the value for the external loss (25%), it corresponded to −2.2 dB of squeezing at the laser output facet. The noise increased at temperatures lower than 140 K. At 80 K, the squeezing reduced to −0.9 dB. At a given temperature, the noise was squeezed only at some values of the pump currents. As the pump current was increased, the noise peak of 1 to 10 dB above the shot-noise value was observed whenever the

Table 6.3. The noise measurement results of the TJS lasers fabricated with different V/III ratios.

Property	V/III ratio	
	156	475
Measurement temperature	140 K	80 K
Threshold current	2.5 mA	1.6 mA
External quantum efficiency (differential quantum efficiency)	0.45	0.54
Measured squeezing	−1.5 dB	−2.8 dB
Corrected squeezing	−2.1 dB	−4.6 dB
$\dfrac{\text{Measured noise}}{\text{Theoretical noise}}$	1.27	1.06

central mode hopped to the next mode. Further increase of the bias stabilized the mode, and the noise again went below the shot noise. This behavior was repeated as the pumping was increased.

In L475 lasers the higher squeezing was observed at the lower temperatures. These lasers exhibited squeezing when the pump current was higher than four times the threshold. The squeezing smoothly increased (normalized noise decreased) with the increase in the pumping level. The noise did not show any jump with the change in the pump current. Although the central mode hopped to the next mode at some operating current, no jump of the noise above the shot noise was observed. Sometimes it caused a small decrease in the squeezing, which was quickly recovered by a slight increase in the pump current. About −2.8 dB of squeezing was measured when the lasers were biased to about 20 times the threshold. After correcting for the loss in the measurement setup, this corresponded to −4.5 dB of squeezing at the laser output facet.

The noise behavior of L475 lasers was reproducible. We characterized nine nominally identical lasers which were cut from the same wafer (Fig. 6.2). These lasers were 200 μm long, 10% AR coated, and mounted on a SiC submount to reduce the strain at low temperatures. The scatter in the noise of different lasers is remarkably small, which demonstrates good control and uniformity in the fabrication process. Figure 6.2 shows −2.5 dB of squeezing at about 20 times the threshold. These lasers showed slightly smaller value of squeezing compared to the value given in Table 6.3 because of the SiC submount. The influence of the mounting material will be discussed in Sect. 6.4.3.

The theoretical value of squeezing (or the normalized noise) far above the threshold is $1-\eta_e$, where η_e is the external QE. For measured values of η_e equal to 0.45 (L156) and 0.54 (L475), expected squeezing was −2.59 dB and −3.37 dB, respectively (Table 6.3). The squeezing observed in the experiments was −1.5 dB and −2.8 dB, respectively. This implies that the experimental squeezing (the reduction in the noise below the shot-noise level)

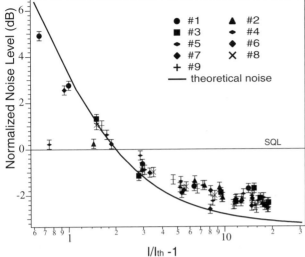

Fig. 6.2. Intensity noise versus pumping (bias current) at 80 K for nine TJS lasers grown with a V/III ratio = 475. The noise was normalized by the corresponding shot-noise value. The line at 0 dB refers to the shot-noise value

Table 6.4. Dependence of noise on AR-coated front-facet reflectivity. Measurements were performed with lasers of 200 μm length that were without the submount.

AR coating	Noise (corrected)	QE (corrected)	I_{th} (mA)
3%	−1.4 (−2.2) dB	0.51 (0.74)	2.35
10%	−2.2 (−3.7) dB	0.49 (0.70)	1.5
35%	−1.8 (−2.9) dB	0.39 (0.57)	0.75

is somewhat smaller than the theoretically expected value. Nonetheless, the noise of L475 lasers was only 1.06 times the expected value.

6.4.2 Optimization of External Coupling Efficiency

In order to achieve higher squeezing from L475 lasers, we tried to enhance the external QE. This was done by reducing the front-facet reflectivity or the length.

Table 6.4 shows that, although the external QE was improved when the AR coating was decreased from 10% to 3%, squeezing was smaller in the latter. Since lasers with a facet reflectivity that is too small are more sensitive to external stray feedback, the higher noise in this case is likely to be caused by optical feedback. In Table 6.4 the amount of squeezing for 10%-coated lasers is less than the value given in Table 6.3 because of slightly higher loss and small polarization sensitivity in this measurement setup. The influence of polarization selection is discussed in Sect. 6.4.3.

The length of the lasers was also reduced to improve the external coupling. In L156 lasers, reducing the length from 200 µm to 150 µm improved the squeezing from -1.5 dB (-2.1 dB corrected for loss) to -2.4 dB (-3.5 dB) and the QE from 0.6 to 0.76. This change agreed with the calculated value, assuming that the reflectivities were same in the two cases. L475 lasers 115 µm long were also made. These lasers did not show significant improvement over 200 µm lasers. We think that it was because the AR-coating reflectivity was higher in the shorter lasers. In our setup, it was difficult to achieve a good AR coating on the short lasers.

6.4.3 Polarization-Partition Noise

The total intensity noise could also be enhanced by the competition between the dominant transverse electric (TE) polarization mode and the weak transverse magnetic (TM) polarization mode. If the suppression of the emission in the TM mode is not enough, it could start lasing. Then, TE and TM modes compete for the common gain and thus each of them carries high polarization-partition noise analogous to the longitudinal-mode-partition noise. In this case any polarization-selective element which transmits one of the two polarization modes more than the other degrades the perfect anticorrelation between the polarization-partition noises of the two modes. The degradation in the anticorrelation between the two polarization modes leads to an increase in the measured noise. The effective gain of the TM mode is influenced by the strain-induced change in the band structure. The excess noise in the QW lasers could have a significant contribution from the imperfect cancellation of the polarization-partition noise, especially when the lasers are cooled to cryogenic temperatures.

In the bulk active medium of a TJS laser, the TE- and TM-mode gains are less sensitive to the strain than in a QW laser. Although the selection of the dominant TE mode by a polarization-sensitive beam splitter did enhance the noise of our TJS lasers, squeezing was not completely destroyed. To investigate the effect of the strain on the noise of the TJS lasers, we tested various laser mounting schemes. Some of the lasers were mounted up-side down, i.e., p-side down, and some with up-side up, i.e., p-side up, on the SiC submount. We also mounted some lasers directly on the copper heat sink without the submount keeping the p-side up.

The thermal-expansion coefficient of SiC (3.7×10^{-6}/deg) is lower than that of either GaAs (6.5×10^{-6}/deg) or copper (17×10^{-6}/deg). The difference in the thermal-expansion coefficients of copper, SiC and GaAs induced different strains in these three mounting schemes when lasers were cooled to 80 K. Because the thermal-expansion coefficient of SiC is lower than that of GaAs, tensile strain was developed in the lasers which were mounted on SiC. On the other hand, compressive strain was developed in the lasers which were

Table 6.5. Dependence of noise on laser mounting. Measurements were performed with lasers of 200 μm length and 10% AR coating. QE: quantum efficiency

p-side up/down (mount material)	Strain	Noise (dB) (corrected)	TE noise (corrected)	QE (corrected)
Up (Cu)	Compressive	−2.8 (−4.5)	−2.2 (−3.2)	0.53 (0.70)
Up (SiC)	Tensile	−2.5 (−3.8)	−1.8 (−2.6)	0.53 (0.70)
Down (SiC)	Tensile	−2.1 (−3.0)	−1.6 (−2.2)	0.53 (0.70)

directly mounted on the copper heat sink due to the higher thermal-expansion coefficient of copper compared to GaAs. The compressive strain reduces the TM-mode gain, while the tensile strain increases it [42]. Table 6.5 presents the total and TE-mode noise of TJS lasers in the three mounting schemes. Even after filtering out the TM mode, −2.2 dB of squeezing was measured for the lasers without the submount. The lasers which were mounted with the p-side down showed the least(−2.1 dB) amount of squeezing, and those with the p-side up without the submount showed the best(−2.8 dB) amount. The p-side down lasers had a larger tensile strain compared to the lasers with the p-side up because the active layer was closer to the mounting interface in the former. The lasers without the submount had a compressive strain and thus a slightly lower gain for the TM mode.

6.4.4 Longitudinal-Mode-Partition Noise

The spectra of these lasers featured highly multi-mode behavior (Fig. 6.3). The first side-mode intensity was only 2 to 5 dB below the central mode intensity even when −4 dB squeezing was achieved. The side modes were reduced to more than 20 dB below the central mode by injection-locking with an external cavity diode laser with narrow linewidth. The noise, however, remained the same as that of the free-running laser (Fig. 6.2). This is in contrast with the reduction of noise by injection-locking observed in QW lasers. This insensitivity of the noise of TJS lasers to the side-mode intensities strongly suggested that the intensity noise correlation among the longitudinal modes was almost perfectly conserved. Thus, the longitudinal-mode-partition noise had a negligible contribution to the total intensity noise of the free-running TJS lasers.

The characteristics of TJS lasers can be attributed to the small amount of saturable loss. A model for the influence of saturable loss on the intensity noise of multi-mode lasers was described in Chap. 5 [142]. The model predicts that nonlinear loss could enhance the total intensity noise when the side-mode intensities are higher. The power in the dominant main mode saturates the nonlinear loss most effectively for the main mode itself. The nonlinear loss for the side modes is not saturated so effectively. This imbalance of the saturable nonlinear loss reduces the anticorrelation among the longitudinal modes. The

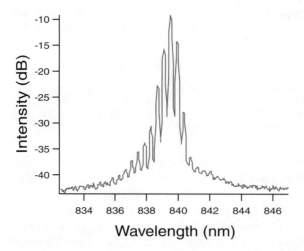

Fig. 6.3. The spectrum of a TJS laser at 80 K and with a 31.1 mA bias current (threshold = 1.5 mA)

physical origin of nonlinear loss considered in the model was the saturable absorption loss by the DX centers in the AlGaAs cladding layers that are created by the donor (Si) impurities. In Chap. 5 we compared the saturable loss due to Si impurity in TJS lasers with that in QW lasers. Because of the smaller saturable loss in TJS lasers, this degradation in the anticorrelation among the longitudinal modes is negligible.

6.4.5 Suppressed 1/f Noise

The noise of semiconductor lasers increases at lower frequencies. This noise, which is approximately inversely proportional to the frequency, is referred to as 1/f noise. The origin of this noise in semiconductor lasers is not well understood. We also measured noise of our TJS lasers in the tens-of-kilohertz range of frequencies (Fig. 6.4). The squeezing was observed at frequencies as low as 40 kHz (Fig. 6.5). The absolute value of the 1/f noise in these lasers was only few dB above the shot-noise level at 100 kHz, which is considerably smaller than for other semiconductor lasers. For comparison, the top trace in Fig. 6.4 plots the noise of a typical QW or TJS laser at room temperature. At room temperature, the noise at 100 kHz is usually in the range of 10 to 20 dB above the shot-noise level. We think that the lower 1/f noise in TJS lasers can be attributed to the smaller density of defects (vacancies and shallow traps).

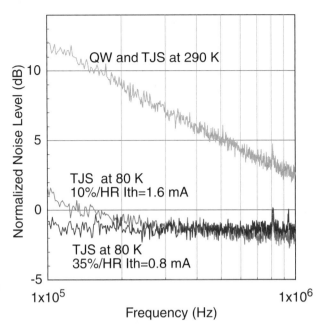

Fig. 6.4. Normalized intensity noise versus frequency for TJS lasers at 80 K (*bottom two traces*) and for a typical TJS or QW laser at room temperature (*top trace*). The *bottom traces* correspond to two TJS lasers with different front facet reflectivity and hence different threshold currents

6.5 Summary

We fabricated GaAs/AlGaAs lasers which reproducibly show about −4.5 dB of squeezing. The key modification made in the processing was the use of a high V/III ratio during growth in order to increase activation of Si atoms as a donor. Consequently, a relatively small concentration of Si was required. The low Si doping level suppressed the intermixing of the GaAs active layer and the AlGaAs cladding layer and decreased the density of trap defects. A heavy p$^+$ doping surrounded by a narrow and shallow p region was created by double Zn diffusion to obtain lateral optical confinement and to increase the carrier-scattering rate. These changes yielded smaller internal loss (linear and saturable), higher internal QE and homogeneous gain broadening. Squeezed light was consistently produced by the new TJS lasers fabricated with optimized cavity design and process conditions.

We also varied the external coupling efficiency by changing the AR coating and the mounting scheme to enhance squeezing. The best squeezing was observed with 10%-AR-coated lasers which were mounted on a copper heat sink without the SiC submount. The lasers with a lower AR coating showed less squeezing than 10%-coated lasers because of increased sensitivity to the small amount of feedback from external optics. About −2.2 dB squeezing

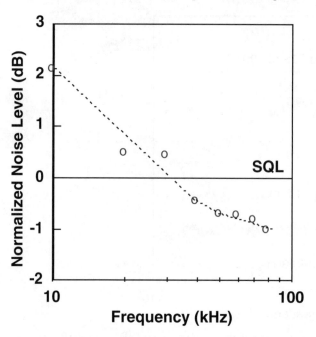

Fig. 6.5. Normalized intensity noise versus frequency at 80 K for a TJS laser with a 35% front-facet reflectivity and a 0.8 mA threshold current. The noise was squeezed at frequencies higher than 40 KHz

(-3.2 dB after correction for the detection loss) was observed in the TE mode of the lasers mounted p-side up and without the submount. Without any polarization selection, a total squeezing of -2.8 dB (-4.5 dB after correction) was measured from the same lasers, which indicated that the polarization mode-partition noise contributed by the TM mode was small.

These TJS lasers can serve as a single-mode, polarized, squeezed light source. TJS lasers under injection-locking may be employed as reliable quasi-single-mode squeezed light sources in applications such as precision spectroscopy and interferometry. The large squeezing in a single polarization would improve the sensitivity in various applications which need polarized light. The low-noise behavior of these TJS lasers at kilohertz frequencies make them particularly attractive for applications as a pump source for diode-laser-pumped solid-state lasers.

7. Sub-Shot-Noise FM Spectroscopy

7.1 Introduction

Frequency modulation (FM) spectroscopy is a very sensitive technique for measuring very small interactions. It was originally proposed as a method for locking a microwave source to a cavity [192] but was later extended to optical carrier frequencies by Bjorklund in the early 1980s [24]. Applications of FM spectroscopy include trace molecular detection, Raman spectroscopy, combustion control, fluorescence spectroscopy, process monitoring, and laser stabilization [41, 86, 148, 153].

There are a number of key features of FM spectroscopy which make it attractive for precision spectroscopy. One critical feature is that measurement is made at radio frequencies (RF), where the amplitude noise spectrum of most lasers is shot-noise limited. Working at RF frequencies is advantageous because the RF electronics industry is very mature, and extremely low-noise electronics are readily available. Finally, in FM spectroscopy the measurement can be made on a zero baseline so the background noise is determined by the intrinsic amplitude noise in the probe laser light at the measurement frequency.

Before the experimental demonstration of photon squeezing, shot noise was considered a fundamental limit to the sensitivity of FM spectroscopy, because lasers were thought to be inherently shot-noise limited. The production of squeezed light was only the first step in making a measurement with non-classical sensitivity [257]. In this chapter experimental techniques developed by Kasapi, Lathi, and Yamamoto [115, 116, 117, 118] for using amplitude-squeezed light from a semiconductor laser to perform FM spectroscopy with a sensitivity greater than that available from a standard shot-noise-limited laser source are described.

7.2 Advantages of Semiconductor Lasers

Amplitude-squeezed semiconductor lasers overcome some of the limitations of sources based on nonlinear interactions. First, the squeezing bandwidth of a semiconductor laser is very large. Sub-shot-noise fluctuations at frequencies as high as 1 GHz have been measured [162]. The large squeezing bandwidth is important because, as Yurke et al. have pointed out, the squeezing at the modulation sideband frequency gives rise to the reduced noise background in the detected signal [275]. One could, in principle, circumvent the small squeezing bandwidth by creating two narrow squeezing bands, one centered at each sideband, but in practice this is very difficult. Consequently, the large squeezing bandwidth of semiconductor lasers allows much greater measurement frequencies than cavity-based vacuum-squeezed sources [120].

Another advantage of using semiconductor lasers for sub-shot-noise FM spectroscopy is that, as we shall show, large FM sidebands can be induced on the squeezed carrier with arbitrarily small residual amplitude modulation (AM) by a combination of dual current modulation and injection-locking. As is well known in FM spectroscopy, residual AM is a major technical problem, because it introduces a measurement offset and prevents the observation of the laser noise floor [25]. The residual AM was suppressed to well below the shot-noise level by using a new technique of simultaneous modulation of semiconductor lasers in a master–slave configuration.

Finally, semiconductor laser systems are very energy efficient (devices with quantum efficiencies greater than 70% are readily available), relatively inexpensive, and available at a variety of useful wavelengths [246]. In contrast, much of the power in a nonlinear squeezing apparatus is lost in the downconversion process. Such experiments also often require large pump lasers and expensive buildup cavities, putting them out of the reach of many labs.

7.3 Signal-to-Noise Ratio (SNR)

The principles of FM spectroscopy have been thoroughly treated in the literature, so we will only summarize the key points here [24, 25].

Figure 7.1 shows the basic processes in FM spectroscopy. When pure FM sidebands (side arrows) are imposed on a carrier (middle arrow) at an optical frequency ω_0 by phase modulation at ω_m and modulation depth M, no AM is detected on the photodetector because the beatnotes between each of the sidebands and the carrier are 180° out of phase and thus cancel each other out.

When the probe beam interacts with a dispersive or absorptive medium, the balance between the sidebands is destroyed. The upper and lower sidebands and the carrier are phase shifted by ϕ_{+1}, ϕ_{-1}, and ϕ_0 respectively and are absorbed by factors of δ_{+1}, δ_{-1}, and δ_0 respectively. The intensity of the light after the interaction is given by [25]

$$
I(t) = \frac{cE_0^2}{8\pi} e^{2\delta_0} \left[1 - M(\delta_{-1} - \delta_{+1}) \cos(\omega_m t) \right. \tag{7.1}
$$
$$
\left. + M(\phi_{+1} + \phi_{-1} - 2\phi_0) \sin(\omega_m t) \right] \quad ,
$$

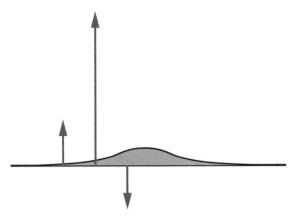

Fig. 7.1. Basic mechanisms of FM spectroscopy. The middle vertical arrow represents carrier and two side arrows represent two sidebands. The sidebands are equal in size and have opposite phase. The shaded region represents a broad spectral line

where M is the sideband amplitude, c is the speed of light, and E_0 is the carrier electric field amplitude. One can separately extract information on the absorption and the dispersion of the medium by mixing the detected signal with the in-phase $\cos(\omega_m t)$ and out-of-phase $\sin(\omega_m t)$ terms.

In our experiments, we measured the total induced RF power by monitoring the amplified RF photocurrent signal on a spectrum analyzer. The square of the mean of the induced photocurrent (which is proportional to the RF power measured by the spectrum analyzer) is given by

$$\langle i_s \rangle^2 = \frac{1}{2} I_0^2 \left(\Delta\delta^2 + \Delta\phi^2 \right) M^2 \quad , \tag{7.2}$$

where $I_0 = e\eta N_0$ is the DC photocurrent, e the electric charge, η the quantum efficiency of the detector, N_0 the average photon flux, $\Delta\delta = \delta_{-1} - \delta_{+1}$ the difference in the absorptions of the two sidebands, and $\Delta\phi = \phi_{+1} + \phi_{-1} - 2\phi_0$ the difference in the phase shifts between the sidebands and the carrier.

The mean square of the photocurrent is proportional to the noise power associated with the RF signal in a bandwidth Δf, which is given by

$$\langle i_n^2 \rangle = \left[2e\eta I_0 \left(1 - \gamma(\omega_m) \right) + 2eI_0\eta(1 - \eta) \right] \Delta f \quad , \tag{7.3}$$

where $\gamma(\omega_m)$ is the squeezing factor at ω_m. $\gamma(\omega_m)$ is defined to be 0 for classical (shot-noise-limited) light and $0 < \gamma(\omega_m) \leq 1$ for nonclassical squeezed light. The signal-to-noise ratio (SNR) is thus given by

$$\frac{\langle i_s \rangle^2}{\langle i_n^2 \rangle} = \frac{I_0 \left(\Delta\delta^2 + \Delta\phi^2 \right) M^2}{4e \left\{ \eta \left[1 - \gamma(\omega_m) \right] + \eta \left(1 - \eta \right) \right\} \Delta f} \quad . \tag{7.4}$$

The first and the second terms of the denominator in (7.4) represent the squeezed amplitude noise of the FM probe wave and the partition noise due to a detector quantum efficiency of less than one respectively.

Amplitude noise squeezing at the measurement frequency improves the SNR because the noise floor is reduced. Of course, reducing the measurement bandwidth would also improve the SNR by decreasing the detected noise power without changing the coherent signal. Unfortunately, reducing the measurement bandwidth comes at the expense of an increased measurement time.

7.4 Realization of Sub-Shot-Noise FM Spectroscopy

In 1987 Yurke et al. showed theoretically that squeezed light could be used to enhance the sensitivity of FM spectroscopy by squeezing the noise in the band around the FM sidebands [275].

In 1992 Polzik et al. demonstrated sub-shot-noise FM spectroscopy using vacuum-squeezed light generated by a degenerate optical parametric oscillator (OPO) [189]. However, a number of technical problems limited the sensitivity of that experiment. The modulation frequency ω_m was only 2.7 MHz in order to make the measurement within the 5.8 MHz half-width at half-maximum (HWHM) of their cavity and hence the squeezing bandwidth. The induced FM sidebands were also only 2% of the carrier amplitude, because larger AM led to an increase in the AM noise. Thus, although their measurement featured a squeezed noise spectrum, the useful measurement frequency bandwidth was severely limited and the signal sensitivity itself was less than optimal because of the small sidebands they were obliged to use.

Our first attempt at sub-shot-noise FM spectroscopy with a squeezed semiconductor laser had problems with achieving good sensitivity [116]. Although the laser amplitude noise floor was reduced below the shot noise, the signal level was low because the induced FM sidebands were very small in order to avoid residual amplitude modulation (RAM). The reduction of RAM was necessary in order to observe the squeezed noise floor, but the measurement SNR was lower than classically predicted value due to limited signal.

For this experiment we developed a new technique for inducing large FM sidebands with negligible RAM. We also improved the squeezing of the laser system from the first experiment by using a transverse-junction-stripe (TJS) laser, described in Chap. 6, instead of the commercial Spectra Diode Labs SDL-5400 series quantum-well laser. We refer to Chapter 6 for a detailed description of the TJS laser used in this experiment.

7.4.1 Frequency and Noise Control by Injection Locking

The linewidth of a free-running semiconductor laser is often very broad because of the low finesse of the cleaved facets. In addition, the anomalous dispersion of semiconductor lasers increases the linewidth above the Schawlow–Townes linewidth by a factor of 10 or more [95]. The central mode full-width at half-maximum (FWHM) in the TJS laser we used in our experiments was over 40 MHz at 140 K. A substantial portion of the total optical power was contained in the longitudinal side modes in these lasers [165]. The first side mode was oscillating with an intensity 20% of the central mode, and hundreds of weaker side modes were observed.

In order to perform FM spectroscopy, the laser linewidth had to be narrowed. The central frequency of the laser also had to be stabilized in order to obtain continuous, controlled tunability of the laser frequency. Over the past decade a number of techniques have been developed to control the frequency spectrum of semiconductor lasers [246]. In particular, injection-locking has proven to be a versatile tool for frequency control of semiconductor lasers [30, 128].

In our experiments we used light from an external-cavity grating-feedback master semiconductor laser to injection-lock a solitary squeezed slave laser. The master laser was a Spectra Diode Labs SDL-5401-G quantum-well semiconductor laser with external feedback from a gold-coated grating [60]. The first-order feedback was approximately 15%. A PZT on the mirror mount holding the grating allowed us to tune the master laser continuously for 6 GHz before a cavity-mode hop occurred.

The free-running spectrum of the TJS laser at 140 K as measured by a scanning Fabry–Perot cavity is shown in Fig. 7.2. The spectrum shows a significant drift in the central oscillation frequency. The free spectral range of the Fabry–Perot cavity was 1.5 GHz, and the finesse was 15 MHz. The free-running optical spectrum measured by an optical spectrum analyzer (OSA) is shown in Fig. 7.3. The resolution of the OSA was 0.1 nm.

As shown in Fig. 7.3, the first sideband was only 5 dB below the carrier, and there were many simultaneously oscillating longitudinal modes. The longitudinal side modes were at or below the threshold and thus individually had an amplitude noise much greater than the shot noise. However, the total laser power was squeezed because of strong negative correlation among the longitudinal side modes. This collective amplitude squeezing, also known as multi-mode squeezing, has been treated elsewhere [62, 105, 106, 165] and is still under investigation.

After injection-locking, the slave laser adopted the single-mode spectral characteristics of the master laser as shown in Fig. 7.4. The spectrum shown in this figure was measured by a scanning Fabry–Perot cavity. The threshold of the slave TJS laser was $I_{th} = 2.25$ mA at $T = 143$ K, with a wavelength of $\lambda = 852.6$ nm, optical injection power $P_{inj} = 1.8$ mW, and total output power $P_{out} = 26$ mW. The longitudinal side modes after injection-locking

Free-Running TJS

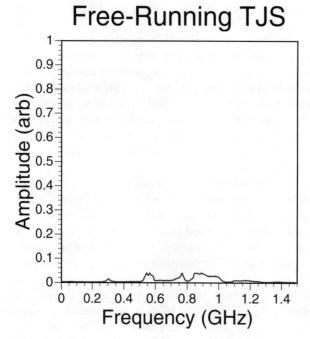

Fig. 7.2. Frequency spectrum of the slave laser before injection-locking, measured by a scanning Fabry–Perot cavity

Free-running TJS

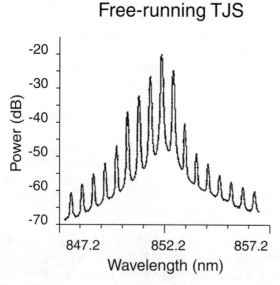

Fig. 7.3. Frequency spectrum of the slave laser before injection-locking, measured by an optical spectrum analyzer

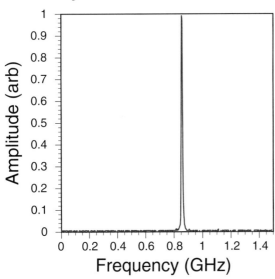

Fig. 7.4. Frequency spectrum of the slave laser after injection-locking, measured by a scanning Fabry–Perot cavity

were measured by an OSA and are shown in Fig. 7.5. The first side modes were 35 dB below the carrier and the other higher-order side modes were suppressed by 10–20 dB.

7.4.2 Effect of Injection Locking on Intensity Noise

The laser noise power and shot-noise level were measured using a standard double-balanced homodyne detector [208]. The sum port hybrid 0/180° mixer measured the laser noise, while the difference port measured the corresponding laser shot-noise level. The shot-noise level was also calibrated with an infrared-filtered fiber-bundle tungsten light source.

Before injection-locking, the total induced photocurrent from the TJS laser used in these experiments was squeezed to 2.0 dB below the shot-noise level. As has already been discussed in Chap. 4, the squeezing was the result of anticorrelation among many weak side modes.

Injection-locking of the laser reduced the strength of the longitudinal side-modes but did not improve squeezing. Injection-locking of several other TJS semiconductor lasers either reduced squeezing or produced no improvement at all. Although the squeezing was slightly less than the theoretical maximum, this less-than-ideal squeezing was probably not caused by mode-partition

Injection-locked TJS

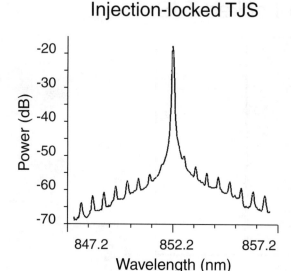

Fig. 7.5. Spectrum of the injection-locked TJS laser, measured by an optical spectrum analyzer

noise among the longitudinal side modes. We are currently investigating other possible mechanisms to account for this excess noise.

7.4.3 Suppression of Residual AM by Injection-Locking

Inducing pure FM is critical to achieving full sensitivity in FM spectroscopy. Reducing the residual AM has been one of the major challenges in demonstrating shot-noise-limited FM spectroscopy. A number of techniques have been developed to reduce the residual AM in a frequency-modulated laser beam [153, 200, 245, 254]. We have developed an all-electrical technique for inducing FM and suppressing RAM using simultaneous current modulation of the master and slave lasers.

FM spectroscopy has been performed using current-modulated semiconductor lasers since the early 1980s [148, 149, 150, 151, 179]. FM sidebands were induced on the output of a semiconductor laser by modulating the pump current [129].

For weak modulation, the amplitude of the sidebands is proportional to the amplitude of the current modulation. However, direct current modulation of a semiconductor laser simultaneously modulates both the index of refraction and the gain of the laser, because the gain and index of refraction are coupled by the parameter α [149, 150, 179]. α describes the ratio between the gain and the index of refraction of the active region when the carrier

Residual AM Reduction

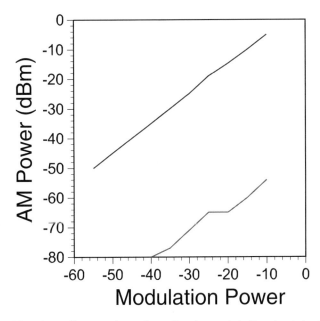

Fig. 7.6. Suppression of amplitude modulation by injection-locking. The upper trace corresponds to the master laser and the lower trace corresponds to the slave laser. The master laser was modulated at 80 MHz.

density is modulated. Previous semiconductor-laser-based FM spectroscopy experiments did not take any measures to suppress the RAM.

Without reduction of the residual AM to well below the laser noise in the measurement bandwidth, the resulting DC offset in the measured FM-to-AM conversion signal would prevent direct observation of the laser noise floor. We have developed a feed-forward current modulation technique for eliminating residual AM. The technique uses simultaneous current modulation and injection-locking of the slave laser by the current-modulated master laser.

The first stage of AM suppression was performed by injection-locking the slave laser with light from a separate master laser. The injection-locked slave adopted the frequency characteristics of the master laser within the locking bandwidth ($\sim$10 GHz). The injection-locked slave frequency was modulated because the master laser featured FM sidebands induced by current modulation. However, as shown in Fig. 7.6, the AM on the injection-locked slave laser was suppressed by over 50 dB below the RAM in the master laser for the same DC photocurrent [48, 116]. The amplitude modulation at 80 MHz

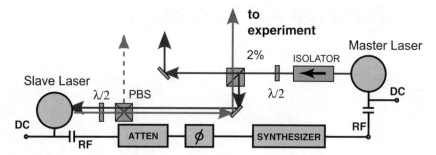

Fig. 7.7. Experimental setup for suppression of residual AM by injection locking and feed-forward modulation. PBS: polarizing beam splitter; RF: radio frequencies

was measured in the master (upper trace) and slave (lower trace) lasers as a function of modulation strength for the same induced photocurrent.

Residual AM suppression occurred because the slave laser was pumped far above the threshold and was thus highly saturated. This step for reducing residual AM alone is a simple and powerful technique for suppressing AM while maintaining FM. Even after injection-locking, however, the AM on the slave laser was still larger than the noise power in the bandwidth of interest. The AM had to be further suppressed before we could demonstrate sub-shot-noise FM spectroscopy.

7.4.4 Suppression of Residual AM
by Dual Pump Current Modulation

Figure 7.7 shows the experimental setup for the feed-forward modulation technique. The basic idea is that the slave and master pump currents are simultaneously modulated so that the AM on slave laser output induced optically by the master laser is exactly cancelled by the AM induced by current modulation of the slave laser. The master laser was modulated by the output from a frequency synthesizer. The master laser induced both FM and AM on the slave laser, but the AM was approximately 50 dB below the master AM.

A small fraction of the power from the synthesizer was phase shifted and attenuated and mixed with the pump current of the slave laser by a bias tee. By carefully adjusting the amplitude and phase of the slave laser modulation, the residual AM induced from the master laser beam was eliminated.

When the slave laser current was modulated without current modulation of the master laser, the FM on the slave laser was suppressed, while, to the first order, the AM was unaffected. As a result, the slave laser was almost purely amplitude modulated.

When the currents of both the master and slave lasers were modulated such that the phase and amplitude of the slave laser AM was equal and oppo-

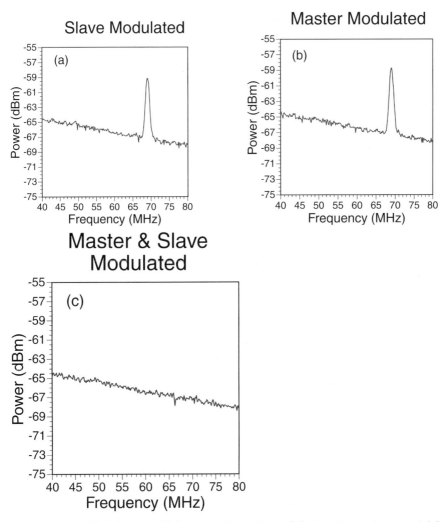

Fig. 7.8 a–c. Modulation of (**a**) master laser alone, (**b**) slave laser alone, and (**c**) both master and slave lasers

site to the AM induced by the master laser, the final AM in the output from the slave laser was suppressed to well below the laser noise level. Figure 7.8 shows the various modulation conditions. Figure 7.9 shows the spectrum of the slave laser measured by a Fabry–Perot cavity with 10% sidebands and feed-forward modulation.

One advantage of this 180° feed-forward modulation technique was that it used only commercially available semiconductor lasers, and no external electro-optic (EO) modulator was required. In fact, EO modulators often

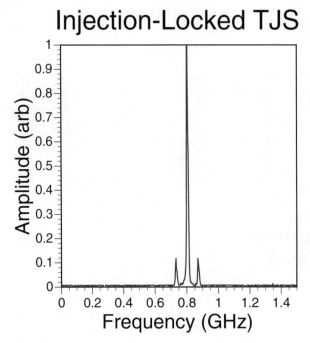

Fig. 7.9. Spectrum of the slave laser with feed-forward modulation, measured by a scanning Fabry–Perot cavity

produce residual AM, and the modulator must be very carefully designed and aligned in order to minimize the residual AM.

One potential limitation to dual pump current feed-forward modulation is that any random fluctuations in the induced AM would prevent cancellation of the AM and introduce noise on the slave laser beam. Under poor injection-locking conditions (such as optical misalignment, tuning the master laser to the edge of the injection-locking bandwidth, or operating the slave laser in an unstable condition), the induced AM was uncorrelated with the RF modulation power on the master laser, and we were unable to suppress the RAM. Under stable operating conditions, however, we were able to reduce the residual AM to well below the background noise level.

Another possible limitation of this technique is that any relative phase fluctuations in the RF modulation signals (such as that induced by propagation down the cable) could cause a phase mismatch between the master laser AM and the opposite AM induced by slave laser current modulation. However, the effect was insignificant in our experiment.

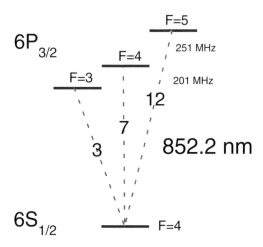

Fig. 7.10. Diagram of the cesium D2 lines

7.4.5 Expected Lineshape

Our atomic sample was thermal cesium in a glass cell with Brewster-angle windows. The modulation frequency was 75 MHz and was limited by the bandwidth of our detector.

The thermally broadened inhomogeneous linewidth of cesium is approximately 500 MHz. In order to calculate the expected lineshape, we averaged over the contributions from the atoms in different velocity classes weighted by the phase density.

The effective Rabi frequency of a two-level atom excited by a laser beam detuned by ω from resonance and with intensity I is

$$\Omega_{\mathrm{R}} = \frac{I}{I_{\mathrm{s}}} \left(\frac{\Gamma}{1 + I/I_{\mathrm{s}} + \omega^2/\Gamma^2} \right) \quad , \tag{7.5}$$

where Γ is the natural linewidth and I_{s} is the saturation intensity (≈ 1 mW/cm^2 for the cesium $|F = 3\rangle \rightarrow |F = 2, 3, 4\rangle$ transitions). This is only the basic interaction for a two-level system. When $I \ll I_{\mathrm{s}}$, the Rabi frequency is linear in the intensity, and thus the absorption is also a linear function of intensity. For cesium there are actually three Doppler-broadened transitions, as shown in Fig. 7.10.

The Maxwell–Boltzmann distribution for a thermal gas at temperature T is

$$P(\boldsymbol{v}) \propto \exp(E/k_{\mathrm{B}}T) \propto \exp[-(v_x^2 + v_y^2 + v_z^2)/k_{\mathrm{B}}T] \quad . \tag{7.6}$$

Integrating $P(\boldsymbol{v})$ over two lateral dimensions v_y and v_z gives the one dimensional Maxwell–Boltzmann distribution

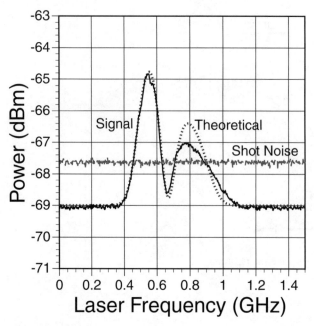

Fig. 7.11. Expected lineshape and experimental full-sensitivity, squeezed FM spectroscopy of thermal cesium

$$P(v_x) \propto \exp(-v_x^2/k_B T) \quad . \tag{7.7}$$

Multiplying $P(v_x)$ by the total number of atoms n gives the number of atoms moving with velocity v_x. Atoms moving with velocity v_x experience a Doppler shift of $\nu v_x/c = v_x/\lambda$, so if the laser is detuned ν from resonance, the resonant atomic velocity class is $v_x = \lambda \nu$ and the number density of atoms as a function of frequency is

$$P(\nu) \propto n \exp\left[(\lambda\nu)^2/k_B T\right] \quad , \tag{7.8}$$

which is also proportional to the beam absorption for $I \ll I_s$. This is the Doppler-broadened line for a single two-level system.

The contributions from the three allowed transitions and the effect of Doppler broadening must all be included in the lineshape. Convolving the allowed transitions with the Doppler width gave the predicted FM-to-AM conversion signal shown as a dotted line in Fig. 7.11. The laser noise as measured in the experiment has also been included in the expected signal to give the -69 dB noise floor.

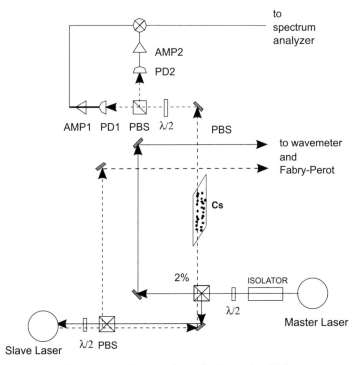

Fig. 7.12. Experimental setup for sub-shot-noise FM spectroscopy with semiconductor lasers. $\lambda/2$: hafl-wave plate; PBS: polarizing beam splitter; PD: photodiode; AMP: amplifier

7.4.6 Spectroscopic Setup

The laser system shown in Fig. 7.7 was used to perform FM spectroscopy on a cesium D2 line. The schematic of the experimental setup is shown in Fig. 7.12.

The slave laser was a solitary TJS GaAs/GaAlAs semiconductor laser designed and produced by our collaborators at Hamamatsu Photonics, Japan. At room temperature the operating wavelength was 890 nm. To reduce the lasing threshold and to tune the wavelength of the slave laser to the cesium 852 nm D2 transition, we cooled and temperature stabilized the laser to approximately 143 ± 0.01 K. At this temperature the free-running slave laser had a squeezed photocurrent 2.1 dB below the standard quantum limit (SQL), with a drive current of 31 mA and threshold of 1.4 mA. At higher drive currents the laser output was chaotic and only excess noise was measured.

To perform FM spectroscopy we passed the squeezed slave laser output through a 4-inch-long quartz spectroscopic cell containing cesium. The windows of the cell were at Brewster's angle to avoid reflection loss. A cold finger on the vapor cell cooled with liquid nitrogen reduced the density of cesium and lowered the signal level to near shot noise. As the laser was swept through the Doppler-broadened $|F = 3\rangle \rightarrow |F = 2, 3, 4\rangle$ transitions, the cesium vapor converted the slave FM into AM. We detected the induced AM by monitoring the sum port of the hybrid mixer in the double-balanced homodyne detector. The shot-noise level was measured on the difference port and double-checked with an infrared-filtered white-light source illuminating the detectors. The sum and difference ports were switched with an electrical RF switch during the spectrum analyzer trace.

We modulated the drive current of the master laser at 75 MHz with a modulation depth of 10% to induce FM sidebands on the master laser. Two percent of the master laser beam was used to injection-lock the cooled slave laser. As described above, after injection-locking, the slave laser only weakly adopted the AM of the master laser but strongly adopted the frequency spectrum. Using the feed-forward technique, the residual AM in the slave laser was reduced to well below the photocurrent noise level for our detection bandwidth of 300 kHz.

7.5 Experimental Results

Figure 7.11 shows an FM spectroscopy trace of thermal cesium which compares favorably to the expected signal, shown as a dotted line. The noise floor was 1.5 dB below the shot-noise level. The asymmetry in the lineshape is caused by the different oscillator strengths for the three cesium D2 transitions from $|F = 4\rangle$. The shot-noise level was measured by monitoring the difference port of the hybrid $0/180°$ mixer in the homodyne detector. The deviation of the measured trace from the theoretical prediction was probably caused by not considering the effect of optical pumping in the theoretical model.

Table 7.1 shows the comparison between our measurement and the earlier sub-shot-noise FM spectroscopy measurement using OPOs [189]. Although that experiment had slightly larger squeezing (3.1 dB versus our 1.5 dB) the modulation frequency was low (2.7 MHz versus our 75 MHz) and the modulation index was small (0.02 versus our 0.1).

The squeezing measured in our apparatus was less than the theoretically expected value. The quantum efficiency of the laser was 65% and the optical efficiency in the setup was 70%, resulting in a theoretical squeezing limit of approximately 3 dB. Several excess noise mechanisms may have contributed

Table 7.1. Comparison to OPO-based sub-shot-noise FM spectroscopy

	Semiconductor Laser	OPO
Modulation frequency	75 MHz	2.7 MHz
Modulation index (M)	0.1	0.02
Squeezing	1.5 dB	3.1 dB

to the increase in noise. The details of these excess noise mechanisms are still an area of active investigation.

7.6 Future Prospects

The production of lasers with reliable squeezing has improved considerably in recent years. The semiconductor lasers we used produced 4.0 dB of squeezing when corrected for optical loss. Without correction we have directly measured 2.5 dB of squeezing in 9 semiconductor lasers which we tested at 77 K without any preselection of the lasers.

As lasers with improved squeezing are developed they will simply replace the slave laser and will immediately lead to further improvement of the SNR. Exactly the same techniques described here can be applied to stabilize the laser frequency and induce FM sidebands. This natural integration of electrical and optical techniques for controlling the spectral properties of squeezed semiconductor lasers paves the way for powerful and simple applications of amplitude squeezing.

With greater squeezing, qualitatively new experiments could potentially be performed. For example, squeezed light could be injected into a cavity to modify the vacuum or could be used to induce nonclassical atom–photon interactions [13, 72, 81, 97]. In addition, injection-locked semiconductor lasers with highly coherent frequency offsets could be used to perform Raman spectroscopy on an amplitude-squeezed noise background.

8. Sub-Shot-Noise FM Noise Spectroscopy

8.1 Introduction

In Chap. 7 we discussed frequency modulation (FM) spectroscopy which required the modulation of the laser frequency. Several groups have recently demonstrated that synthesized stochastic laser fields [4, 54] and the intrinsic, broadband frequency noise of a semiconductor laser [168, 252, 260, 276] can be used to probe resonances in a manner similar to that used in the conventional coherent laser FM spectroscopy. Yabuzaki et al. [260] reported spectra of cesium vapor using diode-laser frequency noise. McIntyre et al. [168] performed direct absorption and pump–probe saturation spectroscopy on rubidium vapor by measuring noise in a narrow bandwidth. Their results resembled those of conventional FM spectroscopy. These experimental results indicate that the frequency noise filtered in a narrow bandwidth can substitute for coherent FM sidebands generated by an external source.

Laser frequency noise stems from spontaneous emission and fluctuations in the refractive index. The intrinsic short photon lifetime and the large anomalous dispersion in semiconductor lasers enhance the frequency noise [95, 238, 261]. This enhanced frequency noise includes a wide laser linewidth with a broad frequency noise pedestal extending out to the relaxation oscillation frequency. For many experiments this excess frequency noise is a serious impediment, because it can cause off-resonant excitations and dephasing of the system being studied.

A constant-current-driven semiconductor laser can have a squeezed amplitude noise spectrum over a bandwidth of several GHz [265], even in the presence of large FM noise. This broadband, large FM noise is an advantage in FM noise spectroscopy.

As in coherent FM spectroscopy, the fundamental background noise in these FM noise experiments is the laser amplitude noise. By combining the large FM noise with the squeezed amplitude noise of a semiconductor laser Kasapi et al. [118] obtained a sub-shot-noise spectrum of laser-cooled rubidium atoms.

In coherent FM spectroscopy, the noise power can be made arbitrarily small by decreasing the measurement bandwidth [25]. The reduction in noise with decreasing measurement bandwidth comes at the expense of an increase

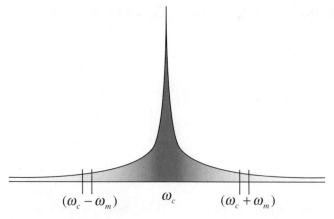

$$(\omega_c - \omega_m) \qquad \omega_c \qquad (\omega_c + \omega_m)$$

Fig. 8.1. Qualitative description of the analogy between coherent FM spectroscopy and FM noise spectroscopy. ω_c is analogous to a carrier wave and $\omega_c \pm \omega_m$ are analogous to modulation sidebands

in the measurement time. In conventional coherent FM spectroscopy, amplitude squeezing is considered an enhancement or convenience which allows faster measurement times.

In FM noise spectroscopy both the noise and the signal strength depend on the measurement bandwidth. Reducing the measurement bandwidth does not increase the signal-to-noise ratio. For lasers with a fixed frequency noise spectrum, reducing the amplitude fluctuations is the only way to increase the signal-to-noise ratio (SNR).

Section 8.4 describes sub-shot-noise FM noise spectroscopy experiment performed on cold rubidium atoms. The measured FM noise spectrum matched the coherent FM spectrum when the absolute value (radio-frequency [RF] power) of the signal was recorded. In this measurement, however, the phase information was lost. Without the phase information, the dispersive (real) and absorptive (imaginary) parts of the probed resonance line cannot be extracted.

In conventional FM spectroscopy the phase information is extracted by mixing the signal with the local oscillator which creates the FM. Such a local oscillator does not exist in FM noise spectroscopy. We demonstrate that phase-sensitive measurements are still possible in FM noise spectroscopy [141]. We extracted the dispersive and absorptive parts of a resonance line using only the frequency noise. Our results matched the theoretical phase-sensitive spectrum expected in conventional FM spectroscopy [25].

8.2 Principle of FM Noise Spectroscopy

FM noise spectroscopy is most easily understood in terms of conventional coherent FM spectroscopy [25]. As described in Chap. 7, a purely frequency-modulated laser beam has two antisymmetric sidebands at $\omega_0 - \omega_m$ and $\omega_0 + \omega_m$ centered at the carrier frequency ω_0. In FM noise spectroscopy the frequency noise in the measurement bandwidth $\Delta\omega_m$ at the measurement frequency ω_m plays the role of the coherent sidebands. Figure 8.1 shows an intuitive model of this frequency noise sideband picture.

When the carrier frequency ω_0 is detuned by ω_m from the atomic resonance, the FM noise sideband at $\omega_0 + \omega_m$ is in resonance with the atomic transition line. The beatnotes between the optical carrier at ω_0 and the two antisymmetric noise sidebands at $\omega_0 - \omega_m$ and $\omega_0 + \omega_m$ no longer cancel each other out, and an electrical FM-to-AM conversion noise signal appears when the optical power is measured, at radio frequency (RF) ω_m on top of the background intrinsic amplitude (AM) noise. Therefore, the same information as that obtained by coherent FM spectroscopy should be available without external modulation of the laser frequency.

8.3 Signal-to-Noise Ratio and the Advantage of Amplitude Squeezing

The SNR for coherent FM spectroscopy was derived in (7.4), and this equation shows that it is inversely proportional to the measurement bandwidth. Increasing the measurement time decreases the measurement bandwidth and thus increases the SNR. However, the improved SNR comes at the expense of a longer measurement time, and eventually drift in the experimental conditions ($1/f$ noise) will become significant.

Another approach for increasing the SNR is by increasing $\gamma(\omega_m)$ above 0, i.e., by using amplitude-squeezed light. Thus, the role of amplitude squeezing in FM spectroscopy is to increase the SNR. Squeezed light has been used to make measurements with noise floors below the shot-noise level of classical coherent states [116, 190]. This idea can also be applied to FM noise spectroscopy. The SNR in FM noise spectroscopy can easily be obtained from the SNR analysis of the coherent FM spectroscopy with only a change in the expression for the signal. The signal strength in the FM noise spectroscopy is given by

$$\langle i_S^2 \rangle = \frac{1}{2} I_0 \left(\Delta\delta^2 + \Delta\phi^2 \right) S_{FM} \Delta f \quad , \tag{8.1}$$

where S_{FM} is the power spectral density of the FM noise (noise power in unit bandwidth) at the measurement frequency ω_m and Δf is the measurement bandwidth. Unlike the coherent signal, the noise signal is proportional to the measurement bandwidth. The SNR is thus

$$\frac{\langle i_S^2 \rangle}{\langle i_n^2 \rangle} = \frac{(\Delta\delta^2 + \Delta\phi^2)S_{\text{FM}}}{4e\left\{\eta\left[1 - \gamma(\omega_m)\right] + (1 - \eta)\right\}} \tag{8.2}$$

and is independent of the measurement bandwidth Δf. Just as in the coherent FM spectroscopy case, the amplitude noise squeezing at the measurement frequency improves the SNR because the noise floor is reduced.

Chapter 7 described sub-shot-noise coherent FM spectroscopy on cesium atoms using a squeezed semiconductor laser system. In this chapter, we present FM noise spectroscopy with a sensitivity greater than that available from a standard shot-noise-limited laser source.

8.4 Sub-Shot-Noise Spectroscopy

8.4.1 Experimental Setup

Figure 8.2 shows the basic elements of the experimental setup. The squeezed semiconductor laser system for FM spectroscopy has been discussed in detail in Chapter 7. We injection-locked the slave semiconductor laser with 2% of the light from another semiconductor laser (SDL-5410-G) stabilized by feedback from an external grating. Injection-locking both controlled the frequency and enhanced the squeezing of the slave semiconductor laser by reducing the strength of longitudinal side modes [106, 242]. The laser system was very similar to that used in our coherent sub-shot-noise FM spectroscopy experiment described in Chap. 7, except that the master laser current was not modulated [116]. As has already been described, external modulation was not necessary, because in FM noise spectroscopy the intrinsic FM noise of the lasers plays the role of the coherent FM sidebands in coherent FM spectroscopy.

The atomic system we studied was laser-cooled rubidium atoms [194]. Rubidium atoms were cooled to $\sim$200 µK in a vapor-loaded magneto-optical neutral rubidium atom trap [74, 173]. We chose this sample because at such low temperatures Doppler broadening is negligible and has a very narrow linewidth of 10 MHz.

The strength of the interaction of the amplitude-squeezed probe light with the sample was controlled by controlling the vapor pressure of rubidium in the trap chamber. A cold finger at $-10°$C on the trap chamber reduced the atomic vapor pressure to minimize the interaction of the probe light with the thermal background rubidium atoms [74]. Interaction with the background rubidium atoms can induce a Doppler-broadened signal and hide the intrinsic laser noise floor.

Light from the slave laser was expanded to approximately 4 cm, $1/e^2$ diameter, to reduce its intensity below the saturation intensity ($\sim$1 mW/cm^2) of rubidium. This expanded beam was used to probe the trapped rubidium atoms. The light could not be simply attenuated because any optical loss

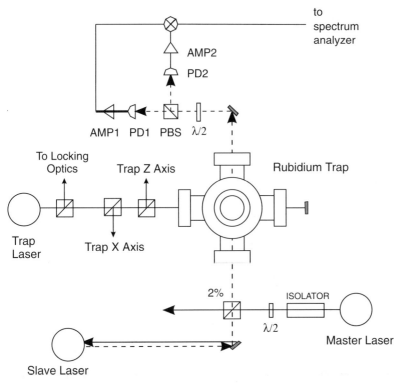

Fig. 8.2. Setup for sub-shot-noise FM noise spectroscopy of magneto-optically trapped rubidium atoms. $\lambda/2$: half-wave plate; PBS: polarizing beam splitter; PD: photodiode; and AMP: RF amplifier

introduces optical partition noise and reduces squeezing. A double-balanced homodyne detector allowed us to measure the laser noise (sum of the two photodetector outputs) and corresponding shot noise (difference of the two photodetector outputs) during a single spectrum analyzer trace.

8.4.2 Laser Trapping and Cooling of Rb

First, we briefly describe the magneto-optical trap used in our experimental setup [45]. Figure 8.3 shows the relevant ^{85}Rb electronic levels for trapping and probing. The trapping laser was an external-cavity semiconductor laser. We used a frequency-offset, saturated-absorption locking scheme in order to produce tunable light at ω_t near the $|5S_{1/2}, F = 3\rangle \rightarrow |5P_{3/2}, F = 4\rangle$ trapping transition. The total output power of the trapping laser was 60 mW, but after frequency-shifting through an acoustic-optical (AO) modulator and subsequent spatial filtering, only 20 mW was available for trapping. Each of the three trapping beams had an intensity of 2 mW/cm^2 and a $1/e^2$ diameter of 7.0 mm. The optical repumping light on the $|5S_{1/2}, F = 2\rangle \rightarrow |5P_{3/2},$

Rubidium 85 (I = 5/2)

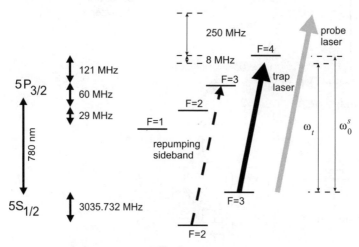

Fig. 8.3. Level diagram for ^{85}Rb

$F = 3\rangle$ transition (needed to avoid populating the $|5S_{1/2}, F = 2\rangle$ hyperfine ground state) was provided by direct-current modulation of the trapping laser at 2.987 GHz to induce a frequency sideband at the repumping wavelength. The trapping magnetic field gradient, provided by two anti–Helmholtz coils wrapped on the outside of the vacuum chamber, was 10 G/cm. From the total trap fluorescence we estimated that the trap contained about 6×10^6 atoms in a ball with a $1/e$ diameter of about 0.5 mm.

8.4.3 Expected Optical Transitions in a Magneto-Optic Trap

Nonlinear spectroscopy of trapped rubidium atoms has been performed by several groups and is relatively well understood [83, 225]. There are three basic resonances in the spectrum of trapped rubidium. The first one is simple absorption of a photon by an atom to excite it from the ground state to one of the excite states (Fig. 8.4). The relevant absorption for our experiment is the Stark-shifted $|5S_{1/2}, F = 3\rangle \rightarrow |5P_{3/2}, F = 4\rangle$ transition at ω_0^s. The second one is stimulated Raman resonance among the Zeeman ground state levels and the trapping photons at the red-detuned trapping frequency ω_t. The third one is the three-photon spontaneous Raman resonance in which absorption of two photons from the trap laser at ω_t and stimulated emission of one photon into the probe laser at $2\omega_t - \omega_0^s$ brings an atom from the ground state to the excited state (Fig. 8.4). The excited state then relaxes to the ground state via spontaneous emission of a photon at ω_0^s.

Among these three, the first and the third resonances were relevant for our experiment because we were detecting the interaction of the probe laser.

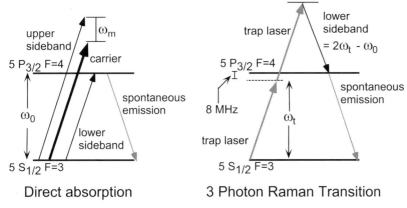

Fig. 8.4. Relevant atomic transitions for trapped rubidium atoms

In the case of frequency-modulated probe, an FM-to-AM conversion signal is generated when one of the FM sidebands comes on to these resonances, because of the difference in the absorption and dispersion of the sidebands.

8.4.4 Sample Probing

When the probe was tuned on to the $|5S_{1/2}, F = 3\rangle \rightarrow |5P_{3/2}, F = 4\rangle$ transition at ω_0^s, rubidium atoms absorbed photons from the intense carrier. The momentum transfer to atoms by the absorbed photons (radiation pressure) was sufficient to overcome the trap potential. Thus, atoms were expelled out of the trap even while the trapping beams and magnetic field were still on. Consequently, the signal became extremely small.

In order to allow time for the atoms to be trapped, the probe laser was detuned 250 MHz to the blue side of the $|5S_{1/2}, F = 3\rangle \rightarrow |5P_{3/2}, F = 4\rangle$ transition at ω_0^s. To make a measurement, the master laser frequency was swept towards resonance ω_0^s in 50 ms by ramping the voltage on a piezoelectric transducer (PZT) controlling the master laser frequency. The injection-locked probe laser exactly matched the frequency of the master laser. The frequency sweep was triggered by the beginning of the time trace of the spectrum analyzer monitoring the output of the double-balanced homodyne detector. The sum of the RF noise signals from the double-balanced detectors was amplified and measured in a 300 kHz bandwidth at a variety of frequencies between 80 and 110 MHz. We chose to measure the noise at a fixed frequency (rather than hold the laser frequency fixed and sweep the measurement frequency) because the response of the detection electronics was frequency dependent.

When the lower sideband approached the resonance at ω_0^s, the carrier was detuned by the measurement frequency (80 to 100 MHz). Even at this detuning the interaction of the intense carrier with the trapped atoms became so strong that the atoms were pushed out of the trap. As has already been

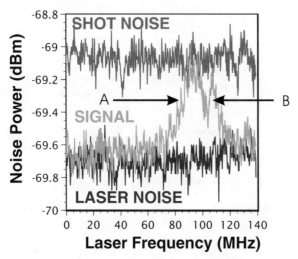

Fig. 8.5. Trace showing sub-shot-noise FM noise spectroscopy data of trapped ^{85}Rb atoms

discussed, enlarging the probe beam to reduce its intensity was crucial in minimizing the mechanical effect due to radiation pressure. We used a repetition rate between 2 and 5 Hz to allow time for the atoms to be re-trapped.

8.4.5 Experimental Result

Figure 8.5 shows the FM-to-AM conversion noise signal without external modulation of the slave laser, as the master laser frequency was swept towards the resonance at ω_0^s. The sample absorption (measured by sweeping a weak probe beam through the sample) was 2%. Our semiconductor laser had a linewidth of about 500 kHz, so assuming a Lorentzian lineshape (valid for the FM noise caused by random phase diffusion), the FM noise spectrum at 90 MHz was about 15 times the AM noise power at the same frequency and in the same measurement bandwidth. This gives an expected FM-to-AM noise conversion signal of 30% the laser amplitude noise. Figure 8.5 shows that the FM-to-AM converted noise power was 20% of the background AM noise power, in rough agreement with the estimated laser noise and sample absorption.

The upper gray trace is the shot-noise level measured by taking the difference between the two photodetector output currents in the double-balanced mixer. The lower gray trace is the intrinsic AM noise of the slave laser plus the FM-to-AM conversion noise due to background ^{85}Rb atoms in the vacuum chamber when the trap was not on. The amplitude noise was squeezed by 0.7 dB below the shot-noise level. The slightly-lower-than-expected signal

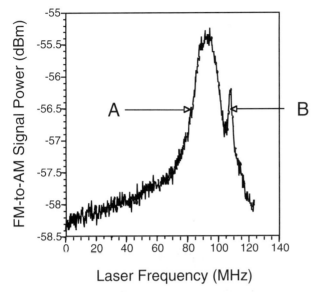

Fig. 8.6. Coherent FM spectroscopy data of magneto-optically trapped ^{85}Rb

level was probably caused by mechanical displacement of the trapped atoms as they absorbed photons from the off-resonant carrier.

Figure 8.6 shows a coherent FM spectroscopy signal taken with a 77.5 MHz external modulation frequency for the purpose of comparing a coherent FM measurement to an FM noise measurement. Peak A corresponds to the Stark-shifted natural absorption of the probe laser FM sideband at ω_0^s and peak B corresponds to the three-photon Raman resonance between the trap laser at ω_t (the trapping frequency) and the probe laser FM sideband. The master laser current was directly modulated with -40 dBm of input RF power, giving an FM modulation index of 2%. The master laser frequency was swept toward resonance at ω_0^s as described in the previous paragraph. The direct absorption by the Stark-shifted $|5S_{1/2}, F = 3\rangle \rightarrow |5P_{3/2}, F = 4\rangle$ transition is clearly visible, as is the Raman transition between the trap and probe lasers. The large background signal was caused by residual AM in the slave laser and by FM-to-AM conversion from the thermal Doppler background rubidium atoms in the trap cell. The shot-noise level was -69.7 dBm and is not visible because of the background signal.

In comparing Figs. 8.5 and 8.6, note that the FM noise spectroscopy signal recovers the same linewidth and spacing as the coherent signal. The signal did not vary greatly with the detection frequency between 80 MHz and 110 MHz. We expected to see a drop in signal level caused by the Lorentzian shape of the noise spectrum. In fact, at higher detection frequencies the carrier was further out of resonance when the noise sideband was inresonance; thus, fewer atoms were mechanically displaced, and the total FM-to-AM noise conversion was stronger.

This experiment demonstrated sub-shot-noise sensitivity in FM noise spectroscopy.

8.5 Phase-Sensitive FM Noise Spectroscopy

The absorption and the dispersion components of a resonance can be measured by homodyne detection, i.e., by mixing the signal and the local oscillator and adjusting the relative phase between them to 0 and $\pi/2$, respectively. Instead of external modulation by a local oscillator, FM noise spectroscopy utilizes intrinsic and stochastic broadband FM of semiconductor lasers. Therefore, there is no external phase reference available for the phase-sensitive homodyne detection.

We generated a reference signal from FM noise by sending part of the semiconductor laser beam through a frequency discriminator. In our experiment the frequency discriminator was a Mach–Zehnder interferometer with unequal path lengths. The frequency discriminator generated amplitude (AM) noise from frequency (FM) noise. Because the FM noise of semiconductor lasers is much larger than the AM noise, this FM-to-AM conversion noise reflects the FM noise in the probe laser beam. Therefore, the FM-to-AM conversion noise signal generated by the discriminator can be used effectively as a reference signal for homodyne measurements. This model is valid for frequencies smaller than the inverse of the delay time between the signal and the reference. In addition, any spurious noise added to either of the beams from optical loss or vibrations of the optics degrades the correlation between the detected probe and the reference signals. This decorrelation, however, can be made very small by using a vibration-isolated table and good optics.

8.5.1 Experimental Setup

Our experimental setup is shown in Fig. 8.7. A quantum-well laser (Spectra Diode Labs SDL-5410-G) was used in a grating-feedback configuration (the first-order back reflection was 15% and the zeroth-order output was 70%). The operating wavelength was 780 nm, and the output beam was divided by a beam splitter.

A Mach–Zehnder interferometer with unequal arm lengths was used as a frequency discriminator to prepare the reference signal. The arm-length difference, ΔL, was set to 24 cm (800 ps). The interferometer output depended on the relative phase, $\Delta\psi$, of the interfering beams. The relative phase $\Delta\psi$ between the two beams for a fixed arm-length difference was modulated by the frequency (ω_{laser}) of the input optical wave and is given by $\Delta\psi = \omega_{\mathrm{laser}} \Delta L / c$

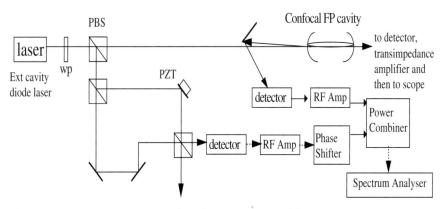

Fig. 8.7. Experimental setup for phase-sensitive FM noise spectroscopy on a Fabry–Perot cavity. Mach–Zehnder interferometer with unequal arm lengths was used as a frequency discriminator to prepare the reference signal. PBS: polarizing beam-splitter, WP: half wave plate; and PZT: piezoelectric transducer

Thus, the frequency fluctuation ($\Delta\omega_{\text{laser}}$) of the input beam resulted in intensity fluctuations in the output when the two arm lengths were not exactly same ($\Delta L \neq 0$).

The output of the Mach–Zehnder interferometer was detected by a p–i–n Si photodiode (Hamamatsu S3394, with area 1 cm^2). The noise in the photocurrent was amplified by two cascaded RF amplifiers (Avantek UTC-517-1, Si microwave monolithic integrated circuit, gain 21.7 dB, cutoff 636 MHz) and then measured by a spectrum analyzer. The measurement frequency was 45 MHz, and the resolution bandwidth (RBW) was 300 kHz. The observed power in this reference signal, E_r, was −39 dBm at the maximum FM-to-AM conversion point (center of an interference fringe). The magnitude of the output depended on the effective path-length difference and measurement frequency.

The probe beam was passed through a confocal Fabry–Perot cavity, which was used to simulate a narrow atomic resonance. The free spectral range and the finesse of the cavity were 1.5 GHz and 200, respectively, and thus the linewidth was 7.5 MHz. Reflection from the cavity was detected by a photodetector (Hamamatsu S3394) and amplified by two RF amplifiers (Avantek UTC-517-1) in cascade.

For phase-sensitive measurement, the reference and probe signals were mixed by a RF power combiner (MiniCircuit ZFC-2-1) and a spectrum analyzer operating at a fixed frequency of 45 MHz. We set the relative phase of the reference and probe signals to 0 and $\pi/2$ by adjusting the length of the delay line used as the phase shifter for reference E_r.

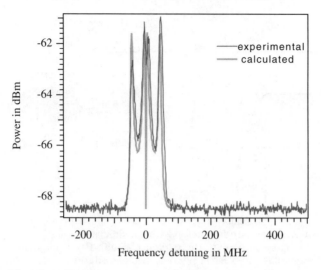

Fig. 8.8. Noise power in the reflection from cavity versus laser detuning. Measurement frequency was 45 MHz, RBW was 300kHz, and photocurrent was 1.5 mA

8.5.2 Experimental Results

We monitored the noise power in the probe signal, E_s, at 45 MHz using a spectrum analyzer as shown in Fig. 8.8. The separation between the noise sidebands of interest was 90 MHz, a value significantly larger than the cavity resonance linewidth 7.5 MHz. Therefore, when the upper (lower) sideband was interacting with the cavity, interaction of the lower (upper) sideband with the cavity was negligible.

The power in the probe signal depends on the relative absorption ($\Delta\delta$) and average dispersion ($\Delta\phi$) of the noise sidebands as given by (7.2). The signal power became maximum when one of the noise sidebands (at ω_m) came into resonance with the line so that the difference between the interaction of the two sidebands was largest. Under this condition the carrier frequency (the laser center frequency), ω_0, was tuned to $\omega_a \pm \omega_m$, where ω_a was the center frequency of the absorption line. When the carrier frequency ω_0 was tuned exactly to the resonance ω_a, both $\Delta\delta$ and $\Delta\phi$ terms became zero. This means that the beat signals of the carrier with the two sidebands canceled each other and the signal power became minimum. Thus, the signal had a double-peaked profile similar to a conventional FM spectrum.

Figures 8.9 and 8.10 show the measurement results when the signal and reference were mixed with relative phase of 0 and $\pi/2$, respectively. These spectrums can be interpreted as explained in the following.

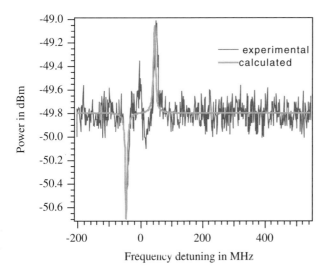

Fig. 8.9. Absorptive component of a Fabry–Perot cavity resonance. This signal was obtained by mixing noise from the beam reflected from the cavity and the beam which traveled through an interferometer, keeping the phase difference between them at 0. Measurement frequency was 45 MHz and RBW was 300 kHz

The RF power generated by this mixing process can be expressed as

$$P_0 = E_{r0}^2 + 2E_{r0}E_{s0}cos(\omega_m t + \varphi)$$
$$\times \ [(\delta_{-1} - \delta_{+1})\cos(\omega_m t) + (\phi_{+1} + \phi_{-1} - 2\phi_0)\sin(\omega_m t)] \quad , \quad (8.3)$$

where E_{r0} is the amplitude of the reference and E_{s0} is the amplitude of the probe signal. Thus, the output signal from the mixer was superimposed upon a background noise level from the reference beam alone (E_{r0}^2).

The $\cos(\omega_m t)$ part of the signal (in-phase-quadrature) was measured by keeping the relative phase between the probe and the reference signals at 0. This in-phase signal was proportional to the difference in absorption of the two noise sidebands. As mentioned above, when one of the sidebands was in resonance with the cavity, the other sideband was detuned far from resonance. Consequently, the difference in the absorption of the two noise sidebands was maximum when the upper noise sidebands came into resonance with the cavity. The peak in Fig. 8.9 corresponds to this frequency. Similarly, when the lower sideband was in resonance with the cavity, the difference in the absorption of the two noise sidebands was minimum (negative and absolute value the same as the maximum). The dip in Fig. 8.9 corresponds to this frequency.

The $\sin(\omega_m t)$ part of the signal (out-of-phase-quadrature) was measured by keeping the relative phase between the probe and the reference signals at $\pi/2$. This out-of-phase signal was proportional to the combined difference in the dispersion of each of the sidebands with the carrier. Since the separation

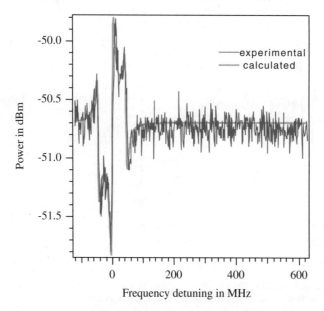

Fig. 8.10. Dispersive component of a Fabry–Perot cavity resonance. This signal was obtained by mixing noise from the beam reflected from the cavity and the beam which traveled through an interferometer, keeping the phase difference between them at $\pi/2$. Measurement frequency was 45 MHz and RBW was 300 kHz

between the carrier and the sidebands was larger than the cavity linewidth, at a given frequency only one of the three (carrier and two sidebands) fields interacted with the cavity. Consequently, the out-of-phase signal was a weighted sum of the individual dispersion of the three fields. The weight was 1 for the sidebands and -2 for the carrier because the out-of-phase signal was proportional to $\Delta\phi = \phi_{+1} + \phi_{-1} - 2\phi_0$. Three dispersive lines are seen in Fig. 8.10. The central feature stems from the dispersion of the cavity imposed upon the carrier. Similarly, two side features arise from the dispersion of each to the two sidebands. The sign of the central dispersive feature is opposite of the two side features.

The extra small peaks in Fig. 8.9 near zero detuning are artifacts, resulting from the fact that the relative phase was not precisely zero. Such imbalance caused partial mixing of the out-of-phase signal (dispersive, Fig. 8.10) with the in-phase signal.

We compared the experimental data with the theoretical spectrum calculated using (8.3) and assuming a Lorentzian lineshape for the Fabry–Perot cavity resonance. The relation given by (8.3) ignores the linewidth of the laser. The nonzero laser linewidth could easily be incorporated by convolving the signal with the lineshape of the laser. The relevant linewidth for our experiment was 300 kHz, the RBW of the spectrum analyzer. This value was much smaller than the cavity linewidth (7.5 MHz), so that the correction due to the nonzero laser linewidth should be very small. The expected signal traces in Figs. 8.8–8.10 show good agreement between experiment and theory.

The asymmetry of the two side peaks in Fig. 8.8 is caused by the intrinsic correlation between the amplitude and frequency noise of the semiconductor laser. As mentioned earlier, the refractive index of the semiconductor medium depends on the carrier concentration (gain); thus, both amplitude and frequency noise are coupled [261]. Therefore, the intrinsic amplitude noise of laser in the measurement bandwidth interferes with the FM-to-AM converted noise signal in that bandwidth. The phase of the FM-to-AM converted signal changes its sign across the resonance. Consequently, the interference of the signal with the amplitude noise is constructive on one side and destructive on the other side of the resonance. Therefore, the two peaks, produced when each of the two sidebands comes into resonance, are of different size.

8.6 Summary

We demonstrated sub-shot-noise FM noise spectroscopy of an atomic rubidium sample using a semiconductor laser. Amplitude squeezing provides a direct improvement of the SNR in any FM noise-based experiment. Indeed, since the SNR is bandwidth independent, reducing the amplitude noise floor is an important refinement of FM noise spectroscopy.

Our sample was magneto-optically trapped and cooled rubidium. The trapped rubidium provided a Doppler-free transition with a rich atomic spectrum dressed by the trapping laser field. We were able to resolve a sub-natural-linewidth Raman resonance as well as the Stark-shifted natural absorption resonance with a signal level entirely below the standard shot-noise level. The observed noise spectrum had qualitatively the same features as the coherent FM spectrum.

We also demonstrated a phase-sensitive homodyne measurement in which the absorption and dispersion components of a resonance were separately measured from the noise signal. The reference signal required in the homodyne measurement was obtained from the laser frequency noise itself. The observed experimental results were in good agreement with theory. This scheme provides a simple method for emulating sensitive FM spectroscopy without any need for expensive high-frequency modulation sources.

Sub-shot-noise FM noise spectroscopy may be useful for probing fragile samples such as trapped cold atoms, optical lattices, and Bose-condensed atoms in which the real absorption of the probe photons must be minimized to avoid back-action on the sample. The interaction can be made very small by using a large modulation frequency and a small modulation index. By the use of large modulation frequency, the strong carrier can be detuned relatively far from the resonance. Thus, the influence of the carrier on the probed system would be small while one of the sidebands probes the line. The small modulation index implies that the amplitude of the sidebands which interacts the sample would also be very small. The resulting signal could still be large for a strong carrier, because the signal is a beatnote between the carrier and the sideband. These features can be easily exploited in FM noise spectroscopy performed with semiconductor lasers, since semiconductor lasers possess intrinsically large and broadband frequency noise.

Additionally, amplitude squeezing reduces the noise floor of such measurements without sacrificing signal strength or measurement bandwidth. The phase-sensitive measurement scheme can be used to extract detailed information about the sample.

We used an external-cavity, grating-feedback semiconductor laser in our experiments, which allows better control of the frequency. The frequency noise of an external-cavity laser is small compared to a free-running laser. The broad linewidth and wide-range tunability of solitary semiconductor lasers make them attractive for such phase-sensitive spectroscopy. However, a solitary semiconductor laser is highly susceptible to weak feedback from optical elements, which can create instability and lead to excess noise [140].

9. Sub-Shot-Noise Interferometry

9.1 Introduction

Amplitude-squeezed light can enhance the sensitivity of interferometric measurements in the same way as it does for spectroscopy. The ultimate limits of a measurement are sought in many fundamental physics experiments which are trying to measure very weak interactions. One example of such difficult but exciting experiments is the detection of gravitational wave which has remained elusive since Einstein predicted the existence of a gravitational wave in general relativity theory. A very-large-scale optical interferometer is now considered as the most promising gravitational wave detector owing to the recent progress of lasers and other optical-component technologies [39]. Inoue, Yamamoto, and Björk proposed a new scheme to realize sub-shot-noise interferometry using amplitude-squeezed light generated directly from a constant-current-driven semiconductor laser [107]. The idea was to inject the amplitude-squeezed light from an injection-locked semiconductor laser into the open port of a dark fringe interferometer. The output noise of the interferometer was reduced to the noise level of the squeezed laser when the interferometer was operated at a dark-fringe.

Since a gravity wave is a very weak disturbance, a long-baseline Michelson interferometer with high-power lasers has been regarded as a promising device for its detection. One important factor besides the laser intensity noise that imposes a practical limit on the sensitivity of such an interferometer is phase modulation of the laser inside the interferometer [213, 65]. Electro-optic modulators put inside the Michelson interferometer are used to move the measurement frequency band to a quiet high-frequency region and hold the interferometer on a dark fringe. This enables a shot-noise (and sub-shot-noise)-limited measurement and recycling of the bright fringe output port back into the interferometer input to achieve better sensitivity through increased power [56, 240]. In a long-baseline gravitational wave detector, however, this internal modulation method causes several problems. First of all, the phase modulators inside the interferometer introduce wavefront distortions which reduce the interferometer contrast. Second, the optical losses due to the absorption and scattering in the Pockels cell as well as the finite aperture limit the achievable recycling gain. Finally, it is not practically easy to fabricate a crystal for large-scale systems which has a large diameter and high-power-density beams.

In order to realize phase detection without Pockels cells inside the interferometer, a method called "pre-modulation" has been proposed [206]. In this method, phase modulated light is injected into an interferometer which has different path lengths in the two arms. However, this path-length imbalance yielded FM/AM conversion noise which is much larger than the shot-noise level in the prototype detector experiment [227]. It would be better not to introduce any path-length difference, to avoid undesired effects caused by the frequency fluctuations.

Inoue and Yamamoto proposed a scheme to realize phase detection not only without Pockels cells inside an interferometer but also without path-length imbalance [110]. The scheme can realize sub-shot-noise sensitivity using an amplitude-squeezed state of light as one of the dual inputs.

In this chapter dual-input interferometer schemes and preliminary experimental results with amplitude-squeezed light injection and external phase modulation are discussed.

9.2 Sensitivity Limit of an Optical Interferometer

The phase-measurement sensitivity limit of an ideal optical interferometer using normal laser light (a coherent state of light) is $\Delta\theta_{\text{SQL}} = 1/\sqrt{N}$, where N is the total number of photons detected in the measurement time. We can improve the sensitivity of interferometric phase measurements by increasing the laser power. However, the use of a high-power laser imposes damage and undesired nonlinear effects on optical components and photodetectors.

The alternative way to improve the sensitivity of interferometric phase measurements is to use "squeezed states of light" together with normal laser light. The sensitivity limit of a usual interferometer is set by a vacuum fluctuation that enters the open port of the interferometer input beam splitter. The straightforward way to improve the sensitivity is to replace a normal vacuum fluctuation with a squeezed vacuum state [40]. We can thus approach the fundamental quantum limit of interferometric phase measurements, $\Delta\theta_{\text{FQL}} = 1/N$. Various sub-shot-noise interferometers have been demonstrated using a squeezed vacuum state generated by nonlinear optical processes [22, 81, 256]. The squeezed vacuum state, which is difficult to realize and use, can be replaced by the amplitude-squeezed light generated from a semiconductor laser. As a prototype for an amplitude- squeezed light interferometer we will first discuss the dual-input Mach–Zehnder interferometer. We will then discuss the dual-input Michelson interferometer, which is particularly promising for gravitation wave detection. The sensitivity limit, however, is independent of the interferometer type.

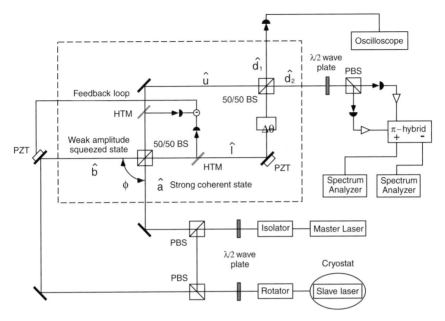

Fig. 9.1. The experimental setup for simultaneously measuring the intensity noise of the interferometer output and the corresponding shot-noise level while the interferometer arm-length difference is changed. BS: beam splitter; PBS: polarization beam splitter; HTM: high- transmission mirror; PZT: piezoelectric transducer

9.3 Amplitude-Squeezed Light Injection in a Dual-Input Mach–Zehnder Interferometer

The experimental setup enclosed in the dashed lines in Fig. 9.1 is a dual-input Mach–Zehnder interferometer. A strong coherent state, $\hat{a}$, and a weak amplitude-squeezed state, $\hat{b}$, are simultaneously injected into the interferometer input beam splitter. The phase differences between the two input fields and that induced by the Mach–Zehnder interferometer arm-length difference are denoted by ϕ and θ, respectively. The operating point of this interferometer is at $\theta = 0$ and only one output field, $\hat{d}_2$, from the second beam splitter is detected. When the weak amplitude-squeezed state is blocked, this interferometer is often referred to as a "dark-fringe interferometer."

The photodetector measures the photon number $\hat{d}_2^\dagger \hat{d}_2$, where we assume that the photodetector has unity quantum efficiency. When we consider the detection of a small modulated phase shift $\Delta\theta$ around $\theta = 0$, the measured photon number $\hat{d}_2^\dagger \hat{d}_2$ can be expanded as

$$\hat{d}_2^\dagger \hat{d}_2 = \hat{b}^\dagger \hat{b} + \frac{i}{2}\left(e^{i\phi}\hat{a}^\dagger \hat{b} - e^{-i\phi}\hat{b}^\dagger \hat{a}\right)\Delta\theta$$

$$= b_0{}^2 + 2b_0\Delta\hat{b}_1 - \left[\left(a_0 b_0 + a_0\Delta\hat{b}_1 + b_0\Delta\hat{a}_1\right)\sin\phi\right]$$

$$+ \left(a_0 \Delta \hat{b}_2 - b_0 \Delta \hat{a}_2 \right) \cos \phi \Big] \Delta \theta \quad , \tag{9.1}$$

where we linearize input field operators $\hat{a}$ $(= a_0 + \Delta \hat{a}_1 + i\Delta \hat{a}_2)$ and $\hat{b}$ $(= b_0 + \Delta \hat{b}_1 + i\Delta \hat{b}_2)$ around the average values a_0 and b_0 [107]. When the phase difference between the two input fields ϕ is stabilized at $\pi/2$, the signal term $(a_0 b_0 \sin \phi)\Delta \theta$ is maximized and the phase-to-amplitude conversion noise $\left(a_0 \Delta \hat{b}_2 - b_0 \Delta \hat{a}_2 \right) \cos \phi$ is minimized. The amplitude-squeezed state $\hat{b}$ can be phase- locked to the coherent state $\hat{a}$ by an injection-locking technique [106, 242]. Since the coherent state and the amplitude-squeezed state have large coherent excitations, the phase difference between the two input fields ϕ is well defined. Therefore, one can keep ϕ at $\pi/2$ for the maximum sensitivity.

The signal power is expressed as

$$(a_0 b_0 \Delta \theta)^2 = N_a N_b (\Delta \theta)^2, \tag{9.2}$$

where $N_a = a_0{}^2$ and $N_b = b_0{}^2$ are the counted photon numbers per measurement time. On the other hand, the noise power is

$$4 b_0{}^2 \langle \Delta \hat{b}_1{}^2 \rangle = \kappa N_b, \tag{9.3}$$

where $\langle \Delta \hat{b}_1{}^2 \rangle = \kappa/4$ and κ is the squeezing factor ($\kappa < 1$ for an amplitude-squeezed state and unity for a coherent state). Since the intensity of the coherent state $\hat{a}$ is chosen to be much stronger than the amplitude-squeezed state $\hat{b}$ ($N_a \gg N_b$), the minimum detectable phase shift is given by

$$\Delta \theta_{\min} = \sqrt{\frac{\kappa}{N_a}} \simeq \sqrt{\frac{\kappa}{N_T}}, \tag{9.4}$$

where N_T is the total number of photons used in the measurement ($N_T = N_a + N_b \simeq N_a$). This sensitivity should be compared to the shot-noise-limited sensitivity, $\Delta \theta_{SQL} = \sqrt{1/N_T}$. We can see that the shot- noise limit is indeed circumvented by using an amplitude-squeezed state as the input field $\hat{b}$ when the photon number of the coherent state is much larger than that of the amplitude-squeezed state. This condition is always satisfied in the interferometric gravitational wave detection experiment.

We can intuitively understand the noise-reduction mechanism in our interferometer using a phasor diagram. The relation between the two input fields $\hat{a}$ and $\hat{b}$ and the two output fields $\hat{u}$ and $\hat{l}$ from the first beam splitter is depicted in Fig. 9.2. The relative phase difference between $\hat{u}$ and $e^{i\Delta \theta} \hat{l}$ is measured by the interferometer. Since the fluctuations in the field $\hat{a}$ are correlated positively in both $\hat{u}$ and $\hat{l}$, they cancel when the two fields are subtracted at the second beam splitter. On the other hand, the fluctuations in the field $\hat{b}$ introduce anticorrelated noise in $\hat{u}$ and $\hat{l}$, so that they add when the two fields are subtracted. Since phase fluctuations in the field $\hat{b}$ are perpendicular to the difference phasor and much smaller than that, they alter only the phase and not the amplitude of the output signal. The background fluctuations for

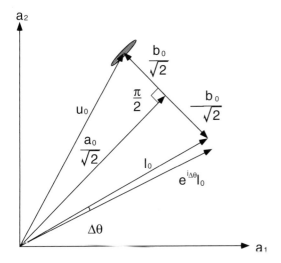

Fig. 9.2. The relation between input fields $\hat{a}$ and $\hat{b}$ and output fields $\hat{u}$ and $\hat{l}$ from the interferometer input beam splitter. a_0, b_0, u_0 and l_0 are the average amplitudes of $\hat{a}$, $\hat{b}$, $\hat{u}$ and $\hat{l}$, respectively

the phase measurement are entirely due to the amplitude fluctuations in the field $\hat{b}$. Therefore, we can improve the sensitivity of interferometric phase measurement using an amplitude-squeezed state as the input field $\hat{b}$.

9.4 Sub-Shot-Noise Phase Measurement

9.4.1 Experimental Procedure

In order to demonstrate the above principle, we simultaneously injected the output from a semiconductor master laser and the amplitude-squeezed state from an injection-locked semiconductor slave laser into a Mach–Zehnder interferometer. In a real gravitational wave detection interferometer, the strong coherent state and the weak amplitude-squeezed state would be the outputs from a high-power Nd:YAG laser and a constant-current-driven low-power In-GaAsP semiconductor laser injection-locked by the YAG laser. In the present experiment, however, both beams are from GaAs semiconductor lasers.

We simultaneously measured the intensity noise of the interferometer output field $\hat{d}_2$ and the corresponding shot-noise value while changing the arm-length difference θ. The master laser is a high-power single quantum well (SQW) semiconductor laser (SDL-5400) which oscillates in a single longitudinal mode with side-mode suppression of about 30 dB. This master laser is mounted in a temperature-controlled chamber and stabilized within several mK. The slave laser is a low-power transverse-junction-stripe (TJS) semi-conductor laser (Mitsubishi ML3401) with an antireflection coating ($\simeq$ 10%) on the front facet and a high-reflection coating ($\simeq$ 90%) on the rear facet. This slave laser is cooled down to approximately 105 K and biased at 20.45 mA. This bias level is 13.6 times the threshold current ($I_{th} \simeq 1.5$ mA), where a single-mode oscillation with side-mode suppression of about 15 dB is obtained. The intensity noise, of the slave laser output is approximately 1 dB below the shot-noise level when free-running and is approximately 2 dB below the shot-noise level when injection-locked. The amplitude squeezing is enhanced with injection-locking by the master laser. This is because the side-mode intensity is suppressed by injection-locking [242, 106] and thus the mode-partition noise which is known to destroy the amplitude squeezing [105, 165] is reduced. A small fraction of the master-laser output power is used for injection-locking, and the rest of the output power (21 mW) is injected into an input port of the interferometer. The amplitude-squeezed state from the injection-locked slave laser (with a power of 14.6 mW) is injected into the remaining open port of the interferometer. Each of the two interferometer arms has a high-transmission mirror ($R \simeq 1$%) which reflects a small fraction of the interferometer beams $\hat{u}$ and $\hat{l}$. The extracted beams are detected by a balanced detector, whose output is used as an error signal to stabilize the phase difference between the two input beams $\hat{a}$ and $\hat{b}$ at $\phi = \pi/2$. One of the outputs from the interferometer, $\hat{d}_1$, is used to monitor the interference fringe. The other output, $\hat{d}_2$, is divided into two beams using a half-wave plate and a polarizing beam splitter and detected by large-area photodiodes (Hamamatsu S3994). The output signals from the photodetectors are amplified and fed into a π-hybrid mixer. The added signal (the intensity noise of the interferometer output) and the subtracted signal (the corresponding shot-noise value) are simultaneously measured using two synchronized RF spectrum analyzers while the arm-length difference is changed.

9.4.2 Experimental Result

Figure 9.3 shows the typical variation of the intensity noise of the interferometer output and the corresponding variation of the shot- noise value as the arm-length difference is changed. The measurement frequency is 10 MHz. We can see that the intensity noise of the interferometer output is reduced to below the corresponding shot-noise value for a certain range of arm- length difference. Within this range, the two arm lengths are almost equal ($\theta \simeq 0$). The maximum intensity noise reduction below the corresponding shot-noise value is 1.58 dB (see Fig. 9.3, inset). The degradation of the amplitude squeezing (approximately 0.42 dB) is due to the interferometer loss (7%) and the imperfect interferences. The visibility of the interference by the master laser output alone (with the slave laser output blocked) is 96.7% and that by the injection-locked slave laser output alone is 96.9%. The imperfect interference by the slave-laser output imposes loss on the amplitude-squeezed state,

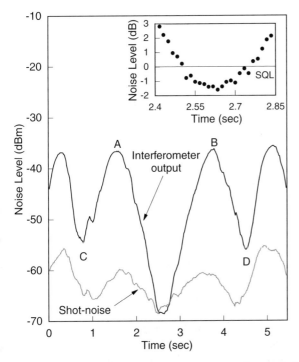

Fig. 9.3. Variation of the intensity noise (sum of the two detector outputs) of the interferometer output (*dark line*) and the corresponding variation of the shot-noise value (difference of the two detector outputs) (*grey line*) while the arm-length difference is changed. The measurement frequency is 10 MHz. The inset shows the normalized intensity noise at around $\theta = 0$. SQL: Standard quantum limit

while the imperfect interference by the master-laser output adds a part of the master-laser output noise to the amplitude-squeezed state. On the other hand, the visibility of the interference between the master-laser output and the injection-locked slave-laser output is only 47.4%. This is because the beam profile of the master-laser output (SQW laser) is markedly different from that of the injection-locked slave-laser output (TJS laser). This imperfect interference between the master-laser output and the slave-laser output does not degrade the amplitude squeezing but reduces the magnitude of the signal power and thus degrades the signal-to-noise ratio.

We can also see the huge excess noise at both sides (points A and B) of $\theta \simeq 0$. When the phase difference between the master- and slave-laser outputs, ϕ, is not stabilized at $\pi/2$, the phase noise $\Delta\hat{b}_2$ of the slave-laser output, which cannot be fully suppressed by injection-locking, is converted into the amplitude noise [see (9.1)]. This phase-to-amplitude conversion noise is maximized at the arm-length difference of $\theta = \pi/2$. The locking bandwidth of the slave laser is approximately 5 GHz, while the cold-cavity bandwidth is considered to be about 1 THz. If we take into account the linewidth enhancement factor $\alpha \simeq 3$ for a TJS laser, the residual phase noise after injection-locking is 56 dB above the shot-noise level. Then the phase error $\Delta\phi$ ($= \phi - \pi/2$) must be $\pm 10°$ to explain the measured excess noise at points A and B.

In Fig. 9.3, we cannot see the intensity noise of the master-laser output (approximately 3 dB above the shot-noise level), which should be measured when the arm-length difference θ is equal to π. At this arm-length difference, the interferometer output should be composed of only the master-laser output. However, the slave-laser output also exists because of the imperfect interference (visibility $\simeq 96.9\%$). There is a possibility that such a residual slave-laser output interferes with the master- laser output. If the interference between the master-laser output and the residual slave-laser output is strong, the phase noise of the residual slave- laser output is homodyne-detected by the intense master-laser output, as shown in Fig. 9.4a. This homodyne-detected phase noise increases the intensity noise of the interferometer output at $\theta = \pm\pi$ (points C and D).

The same is true at the operating point of this interferometer. When the arm-length difference θ is equal to zero, the interferometer output should be composed of only the slave-laser output. However, the master–laser output also exists because of the imperfect interference (visibility $\simeq 96.7\%$). If the interference between the slave-laser output and the residual master- laser output is strong, the residual phase noise of the slave-laser output is homodyne-detected by the residual master- laser output as shown in Fig. 9.4b. The intensity noise of the interferometer output at $\theta = 0$ is shown in Fig. 9.4c as a function of the strength of the interference between the slave-laser output and the residual master- laser output. Here, we assume that the phase error $\Delta\phi$ is $\pm 10°$, and the residual phase noise of the injection-locked slave-laser output is 56 dB above the shot-noise level. If the slave-laser output interferes with the residual master- laser output, then the squeezed intensity noise of the interferometer output is easily destroyed by the homodyne-detected slave-laser phase noise. In order to realize sub-shot-noise interferometry, it is therefore important to reduce the residual phase noise of the injection-locked slave-laser output, and obtain high visibilities.

The difference between the maximum photocurrent and the minimum photocurrent is 4.5 dB, corresponding to a visibility of 47.4%. The difference between the maximum and the minimum noise values (grey line in Fig. 9.3) measured by the subtracted port of the π- hybrid is more than 4.5 dB. This is due to the imperfect common-mode suppression of our detection system. The common-mode suppression of the π- hybrid we used in this measurement is approximately 30 dB.

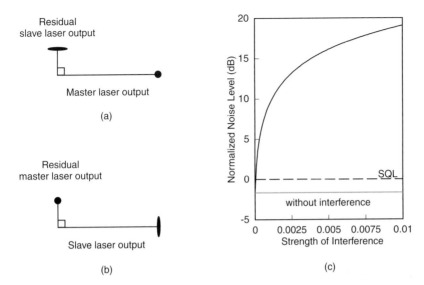

Residual
slave laser output

Master laser output

(a)

Residual
master laser output

Slave laser output

(b)

(c)

Fig. 9.4 a–c. Homodyne-detected residual phase noise of the injection-locked slave-laser output. (**a**) $\theta = \pi$, (**b**) $\theta = 0$. (**c**) The normalized intensity noise of the interferometer output at $\theta = 0$ as a function of the strength of the interference between the slave-laser output and the residual master-laser output. The *grey line* shows the normalized intensity noise when the slave-laser output does not interfere with the residual master-laser output. SQL: Standard quantum limit

9.5 Dual-Input Michelson Interferometer

9.5.1 Operation Principle

The dual-input Michelson interferometer is depicted in Fig. 9.5. High-power laser light is divided into two beams using a high-transmission mirror. The

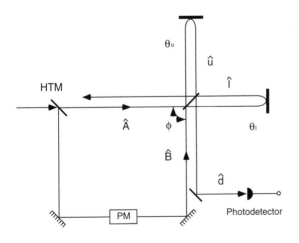

Fig. 9.5. Schematic layout of a dual-input Michelson interferometer. HTM: High-transmission mirror; PM: Phase modulator

reflected weak beam is phase modulated using an electro-optic modulator and injected into the port which is usually unused. The phase difference between the transmitted strong beam and the reflected weak beam is ϕ, and the phase shifts in each arm due to the gravitational wave are θ_u and θ_l. Here the laser frequency is ω_0 and the modulation frequency is ω_m. In the following calculation, we assume that the phase-modulation index ϵ is small enough so that $J_0(\epsilon) \simeq 1$ and $J_{\pm 1}(\epsilon) \simeq \pm\epsilon/2$, where $J_0(\epsilon)$ and $J_{\pm 1}(\epsilon)$ are the zeroth and first-order Bessel functions, respectively.

We define field operators for the dual inputs $\hat{A}$ and $\hat{B}$ as follows:

$$\hat{A} = \hat{a}e^{-i\omega_0 t} + \hat{c}_1 e^{-i(\omega_0+\omega_m)t} + \hat{c}_{-1}e^{-i(\omega_0-\omega_m)t} \quad , \tag{9.5}$$

$$\hat{B} = \hat{b}e^{-i\omega_0 t} + \hat{d}_1 e^{-i(\omega_0+\omega_m)t} + \hat{d}_{-1}e^{-i(\omega_0-\omega_m)t} \quad , \tag{9.6}$$

where $\hat{a}$ and $\hat{b}$ are field operators for the transmitted strong beam and the reflected weak beam, respectively. The vacuum fluctuations at sideband frequencies $\hat{c}_1$ and $\hat{c}_{-1}$ in $\hat{A}$ as well as the fluctuations $\hat{d}_1$ and $\hat{d}_{-1}$ in $\hat{B}$ are included in order to see the effect of the phase modulation on the sensitivity of this interferometer. The field operators propagating in each arm are expressed as

$$\hat{u} = \frac{1}{\sqrt{2}}\left(\hat{A}e^{i\phi} + \hat{B}\right)e^{i\theta_u} \tag{9.7}$$

and

$$\hat{l} = \frac{1}{\sqrt{2}}\left(\hat{A}e^{i\phi} - \hat{B}\right)e^{i\theta_l} \quad . \tag{9.8}$$

The output field $\hat{d}$ is expressed as

$$\hat{d} = \frac{1}{\sqrt{2}}\left(\hat{u} - \hat{l}\right) = e^{i\frac{\theta_u+\theta_l}{2}}\left[i\sin\left(\frac{\theta}{2}\right)\hat{A}e^{i\phi} + \cos\left(\frac{\theta}{2}\right)\hat{B}\right] \quad , \tag{9.9}$$

where $\theta = \theta_u - \theta_l$. The photodetector measures the photon number $\hat{d}^\dagger\hat{d}$, where we assume that the photodetector has unity quantum efficiency. The measured photon number $\hat{d}^\dagger\hat{d}$ is given by

$$\hat{d}^\dagger\hat{d} = \frac{1}{2}(1-\cos\theta)\hat{A}^\dagger\hat{A} + \frac{1}{2}(1+\cos\theta)\hat{B}^\dagger\hat{B}$$
$$- \frac{i}{2}\sin\theta\left(e^{-i\phi}\hat{A}^\dagger\hat{B} - e^{i\phi}\hat{B}^\dagger\hat{A}\right) \quad . \tag{9.10}$$

We linearize field operators as follows:

$$\hat{a} = a_0 + \Delta\hat{a}_1 + i\Delta\hat{a}_2 \quad , \tag{9.11}$$

$$\hat{b} = b_0 + \Delta\hat{b}_1 + i\Delta\hat{b}_2 \quad , \tag{9.12}$$

$$\hat{c}_1 = \Delta\hat{c}_{11} + i\Delta\hat{c}_{12} \quad , \tag{9.13}$$

$$\hat{c}_{-1} = \Delta\hat{c}_{-11} + i\Delta\hat{c}_{-12} \quad , \tag{9.14}$$

$$\hat{d}_1 = -\frac{\epsilon}{2}b_0 + \Delta\hat{d}_{11} + i\Delta\hat{d}_{12} \quad , \tag{9.15}$$

$$\hat{d}_{-1} = \frac{\epsilon}{2}b_0 + \Delta\hat{d}_{-11} + i\Delta\hat{d}_{-12} \quad , \tag{9.16}$$

where a_0 and b_0 are the average amplitudes of the transmitted strong beam and the reflected weak beam, respectively. $\Delta\hat{a}_1$, $\Delta\hat{b}_1$, $\Delta\hat{c}_{11}$, $\Delta\hat{c}_{-11}$, $\Delta\hat{d}_{11}$, and $\Delta\hat{d}_{-11}$ are the in-phase fluctuation operators. $\Delta\hat{a}_2$, $\Delta\hat{b}_2$, $\Delta\hat{c}_{12}$, $\Delta\hat{c}_{-12}$, $\Delta\hat{d}_{12}$, and $\Delta\hat{d}_{-12}$ are the quadrature-phase fluctuation operators. We substitute (9.11–16) into (9.10) and neglect fluctuation operators (we will discuss the effect of the fluctuation operators in Sect. 9.5.2) to get the averaged interferometer output

$$I = \frac{1}{2}(1 - \cos\theta)a_0{}^2 + \frac{1}{2}(1 + \cos\theta)b_0{}^2 - a_0 b_0 \sin\theta\sin\phi$$
$$+ \epsilon a_0 b_0 \sin\theta\cos\phi\sin\omega_m t \quad . \tag{9.17}$$

We choose the operating point of this dual-input interferometer to be θ_0 ($\theta_0 \ll 1$). The signal due to the gravitational wave, $\Delta\theta$, is included in the third and fourth terms. The third term is maximized and the fourth term is minimized when the phase difference ϕ between a_0 and b_0 is $\pi/2$. Therefore, we can stabilize the phase difference ϕ at $\pi/2$ using the fourth term to get an error signal and extract the small phase shift due to the gravitational wave, $\Delta\theta$, from the third term. When $\theta = \theta_0 + \Delta\theta \ll 1$ and $\phi = \pi/2$, (9.17) can be expressed as

$$I = b_0{}^2 - a_0 b_0(\theta_0 + \Delta\theta) \quad . \tag{9.18}$$

Therefore, in the dual-input Michelson interferometer, we detect the weak beam $\hat{b}$ and the interference term, while the strong beam $\hat{a}$ can be recycled by putting a recycling mirror just after the high-transmission mirror.

9.5.2 Sensitivity of a Dual-Input Michelson Interferometer

As we have seen in Sect. 9.5.1, the gravitational wave signal appears at DC in a dual-input Michelson interferometer. Therefore, we need to consider the photon-number fluctuation at DC in order to calculate the sensitivity of the interferomenter. The photon-number fluctuation at DC is given by

$$\Delta\hat{I} = 2b_0\Delta\hat{b}_1 - \epsilon b_0 \left(\Delta\hat{d}_{11} - \Delta\hat{d}_{-11}\right)$$
$$- \theta_0 \left[a_0\Delta\hat{b}_1 + b_0\Delta\hat{a}_1 - \frac{\epsilon}{2}b_0(\Delta\hat{c}_{11} - \Delta\hat{c}_{-11})\right] \quad , \tag{9.19}$$

where we assume that $\theta_0 + \Delta\theta \ll 1$ and $\phi = \pi/2$. The second term is much smaller than the first term because we assumed that $J_0(\epsilon) \simeq 1$. In the square bracket, the first term is much larger than the second and the third terms because $a_0 \gg b_0$. Therefore, the photon-number fluctuation at DC is simplified as

$$\Delta \hat{I} = 2b_0 \Delta \hat{b}_1 - \theta_0 a_0 \Delta \hat{b}_1 \quad . \tag{9.20}$$

When we set the phase offset $\theta_0 \ll 2b_0/a_0$, the background noise in the phase measurement is determined by the photon-number fluctuation of the weak beam $2b_0 \Delta \hat{b}_1$.

The signal power is given by

$$(a_0 b_0 \Delta \theta)^2 = N_a N_b (\Delta \theta)^2 \quad , \tag{9.21}$$

where $N_a = a_0{}^2$ and $N_b = b_0{}^2$ are the counted photon numbers per measurement time interval. On the other hand, the noise power is given by

$$4b_0{}^2 \langle \Delta \hat{b}_1{}^2 \rangle = \kappa N_b \quad , \tag{9.22}$$

where $\langle \Delta \hat{b}_1{}^2 \rangle = \kappa/4$ and κ is the squeezing parameter ($\kappa < 1$ for an amplitude-squeezed state and unity for a coherent state).

In conventional modulation methods, the gravitational wave signal appears at a modulation frequency ($\simeq 10$ MHz) and we can avoid the excess laser intensity noise at low frequency. In the dual-input interferometer, the gravitational wave signal appears at DC. However, as long as the intensity noise of the weak beam $\hat{b}$ is shot-noise limited we can get the shot-noise-limited sensitivity. As shown in Fig. 9.5, the weak beam $\hat{b}$ is highly attenuated by a high-transmission mirror (the attenuation can be more than 30 dB). Therefore, the intensity noise of the highly attenuated beam is shot-noise limited even at low measurement frequencies ($\simeq$ kHz). Hence, the minimum detectable phase shift of the dual input interferometer, which is defined by a unity signal-to-noise ratio, is given by

$$\Delta \theta_{\min} = \sqrt{\frac{1}{N_a}}. \tag{9.23}$$

The sensitivity of the dual-input Michelson interferometer is determined by the photon number of the strong beam and is indeed shot-noise limited.

9.5.3 Sub-Shot-Noise Interferometry

In the Sect. 9.5.2, we concluded that the sensitivity of the dual-input Michelson interferometer is determined by the intensity noise of the weak beam $\hat{b}$ when the phase offset θ_0 is set appropriately. In this section, we consider the application of an amplitude-squeezed state of light to the dual-input Michelson interferometer.

Figure 9.6 shows the configuration of the sub-shot-noise interferometer. The output from a Nd:YAG laser (master laser) is divided into two beams

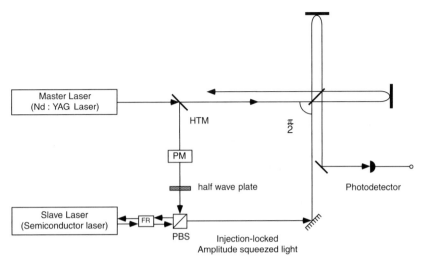

Fig. 9.6. Schematic of a sub-shot-noise dual-input Michelson interferometer. HTM: High-transmission mirror, PM: phase modulator, PBS: Polarizing beam splitter, FR: Faraday rotator

using a high-transmission mirror. The reflected weak beam is phase modulated and injected into a constant-current-driven semiconductor laser (slave laser), which generates an amplitude-squeezed state of light [106, 242]. By injection-locking, identical phase-modulation sidebands are generated as the input signal on the amplitude-squeezed state, while the spurious amplitude-modulation sidebands introduced by the phase modulator are suppressed by the gain saturation of the slave laser [116]. The injection-locked and phase-modulated amplitude-squeezed state of light is then injected into the open port of the beam splitter. Thus, the shot-noise-limited weak beam in the previous section is replaced with the amplitude-squeezed state. Then, the minimum detectable phase shift for this configuration is given by

$$\Delta\theta_{\min} = \sqrt{\frac{\kappa}{N_{\mathrm{a}}}} \quad , \tag{9.24}$$

where $\kappa < 1$. Thus, the required intensity of the strong laser beam inside the interferometer can be reduced while the sensitivity is kept constant.

In conventional sub-shot-noise interferometry using a squeezed vacuum state as one of the dual inputs [40], the gravitational wave signal appears at a modulation frequency and is proportional to the modulation index ϵ. The beat noise between the phase-modulation sidebands of the main beam and the fluctuations of the incident fields from the open port at frequencies ω_0 and $\omega_0 \pm 2\omega_{\mathrm{m}}$ must all be reduced below the vacuum fluctuation level using a broadband squeezed vacuum state to obtain sub-shot-noise sensitivity [70]. In the dual-input interferometer proposed here, the intensity noise of the weak beam needs to be reduced below the shot-noise level only at the measurement frequency to get sub- shot-noise sensitivity, since the gravitational wave signal appears at DC.

9.6 Summary and Future Prospects

Development of highly amplitude-squeezed and good single-mode lasers would allow very sensitive phase measurements. The proposed sub-shot-noise interferometry will be incorporated into the dark-fringe gravitational wave detection interferometer if the residual phase noise of the injection-locked slave laser is suppressed to an appropriate level.

In an actual gravitational wave detection interferometer, spatial coherence and good mode matching between a high-power master laser, e.g. Nd:YAG laser, and a squeezed semiconductor laser are required so that interference fringe visibility is high. The Nd:YAG beam profile is circular and the semiconductor laser beam profile is usually elliptical. Thus, beam-profile modification will be necessary.

In a new scheme for interferometric gravitational wave detection using a dual- input Michelson interferometer, a phase modulator can be placed outside the interferometer and the two arm lengths can be balanced. This setup increases the interferometer contrast and the recycling gain, and avoids FM/AM conversion noise due to arm-length imbalance. Since a weak beam injected into the usually unused port of the beam splitter is phase modulated, a commercial electro-optic modulator (which works for low-power light) can be used.

10. Coulomb Blockade Effect in Mesoscopic p–n Junctions

10.1 Introduction

The previous chapters of this book concentrated on the intensity noise suppression of light generated by a p–n junction light emitter in the macroscopic limit, where the charging energy due to a single carrier traversing the depletion region of the junction is negligible compared to the characteristic thermal energy of the system. With today's semiconductor fabrication and cryogenic technologies, it is possible to reach the opposite regime where the single carrier charging energy, e^2/C, is larger than the thermal energy, $k_B T$. The next two chapters will describe p–n junctions operating in this mesoscopic limit, where a single-carrier injection event completely suppresses the rate for a subsequent carrier-injection event. Such limit was first discussed in the context of mesoscopic M–I–M junctions [10, 11], and is commonly referred to as Coulomb blockade effect [79].

This chapter describes a simple model for the carrier dynamics in a mesoscopic p–n junction. Unlike the situation for the macroscopic limit discussed in Chap. 2, (2.11) which describes the junction voltage dynamics cannot be linearized, and the stochastic dynamics need to be treated by Monte-Carlo simulations. A mesoscopic p–i–n junction with an (undoped) intrinsic quantum well (QW) active layer is considered, where carrier injection is achieved via resonant tunneling through tunnel barriers from both n-type and p-type layers. The microscopic model used to calculate the tunneling rates in such a structure is discussed. The effect of Coulomb charging energy on the tunneling rates of carriers can be modeled from such microscopic considerations. Qualitative analysis on this model predicts observation of Coulomb staircase in the DC I–V characteristics of such a small p–i–n junction [21]. When one considers the case where the junction is driven with an AC square modulation voltage with an appropriate frequency, it can be shown that the intensity fluctuation of the light output from such a p–i–n junction can be suppressed to below the Poissonian limit, even at the single photon level [104]. This is in sharp contrast with the experimental situation described in Chap. 3, where the intensity fluctuation of photons was suppressed only for long measurement time scales that involved $N = 10^7 \sim 10^8$ photons per measurement time slot. This opens up the possibility to generate regulated single photons with well-defined time separation, which will be discussed in Chap. 11.

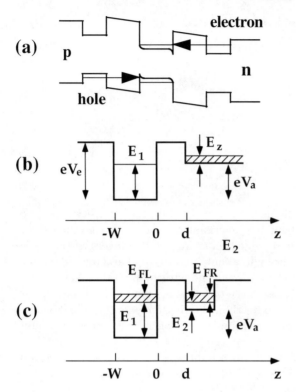

Fig. 10.1 a–c. (a) Schematic energy-band diagram of the p–i–n junction structure under consideration. (b) Parameters for the tunneling matrix element calculation. (c) In the real device, the n-type lead is replaced with a second quantum well (QW)

10.2 Calculation of Resonant Tunneling Rates

Figure 10.1a shows the schematic band diagram of the mesoscopic double barrier p–i–n junction under consideration. The active region is an intrinsic GaAs QW (central QW) in the middle of a p–n junction. Electrons and holes are supplied to the central QW from side QWs on the n-side and p-side, respectively, via resonant tunneling across a tunnel barrier. The n-side and p-side QWs are not deliberately doped, but carriers are supplied from nearby n-type and p-type layers by modulation doping. The lateral size of the device is assumed to be made small, in order to increase the single charging energy $e^2/2C_i$, where C_i (i = n or p) is the capacitance between the central QW and the i-side QW.

The resonant tunneling rates of the electrons and holes from the side QWs to the central QW can be calculated as follows [36]: First, the tunnel barrier as shown in Fig. 10.1b is considered, and the transmittance of an electron (hole) state in the QW to a state outside the well (with the same energy) through this tunnel barrier is calculated, as a function of the energy of the state. Then the escape rate of an electron in this state (in the QW) can be calculated from the transmittance. The expressions for the escape rate can also be calculated if one knows the tunneling matrix element: by equating these two expression for the escape rates, one can find an explicit expression for the tunneling matrix element. Once the tunneling matrix element is known, the tunneling rate of electrons (holes) into the central QW state at a given bias voltage can be calculated by integrating the product of the tunneling matrix element, the Fermi function, and the density of states.

10.2.1 Transmittance of the Barrier

The transmittance of the tunnel barrier shown in Fig. 10.1b is first calculated for electrons by considering an electron escaping from the QW. The height of the barrier, the quantized energy level of the QW, the conduction-band minimum of the n-type lead, and the (in-plane) kinetic energy of the electron in the n-type lead are denoted by eV_e, E_1, eV_a, and E_z, respectively. The width of the QW and the tunnel barrier are W and d respectively, and the origin of the z axis is chosen to be at the interface between the QW and the tunnel barrier. The electron wavefunction inside the QW (Ψ_I), in the barrier (Ψ_{II}), and in the n-type lead (Ψ_{III}) are given by

$$\Psi_I = A\exp(ik_1 z) + B\exp(-ik_1 z) \quad ,$$
$$\Psi_{II} = C\exp(\kappa_1 z) + D\exp(-\kappa_1 z) \quad , \tag{10.1}$$
$$\Psi_{III} = E\exp[ik_2(z-d)] \quad ,$$

with

$$k_1^2 = \frac{2m_I}{\hbar^2}E_1 \quad ,$$
$$\kappa_1^2 = \frac{2m_{II}}{\hbar^2}(eV_e - E_1) \quad , \tag{10.2}$$
$$k_2^2 = \frac{2m_{III}}{\hbar^2}E_x \quad .$$

Although the materials used for the QW, the barriers and the n-type lead are different in our experimental situation, the difference in the effective masses is small and will be neglected in the following discussion ($m_I = m_{II} = m_{III} \equiv m^*$). The boundary conditions that the wavefunction and its first derivative with respect to z are continuous at $z = 0$ and $z = d$ uniquely determine A and B as a function of E. The results are

$$A = \frac{1}{2}\left[\left(1 + \frac{k_2}{k_1}\right)\cosh \kappa_1 d + i\left(\frac{\kappa_1}{k_1} - \frac{k_2}{\kappa_1}\right)\sinh \kappa_1 d\right] E,$$

$$B = \frac{1}{2}\left[\left(1 - \frac{k_2}{k_1}\right)\cosh \kappa_1 d - i\left(\frac{\kappa_1}{k_1} + \frac{k_2}{\kappa_1}\right)\sinh \kappa_1 d\right] E \quad . \quad (10.3)$$

A denotes the amplitude of the wave in the QW propagating in the $+z$ direction, and E the amplitude of the wave in the n-type lead propagating away from the QW. The transmittance of the barrier is given by the ratio of the electron flux incident at the barrier and the flux transmitted through the barrier. The electron flux is given by the current-density operator

$$J_z = \frac{\hbar}{2m^*i}\left(\Psi^*\frac{d\Psi}{dz} - \Psi\frac{d\Psi^*}{dz}\right) \quad . \quad (10.4)$$

The transmittance of the barrier is given by

$$T = J_{\mathrm{III},+z}/J_{\mathrm{I},+z} = \frac{k_2}{k_1}\frac{|E|^2}{|A|^2} \quad . \quad (10.5)$$

In the limit of low transmission ($\kappa_1 d \gg 1$), one can make the approximations

$$\sinh \kappa_1 d \simeq \cosh \kappa_1 d \simeq \exp(\kappa_1 d) \quad , \quad (10.6)$$

and the transmittance is given by

$$T = \frac{4k_1 k_2 \kappa_1^2}{(\kappa_1^2 + k_1^2)(\kappa_1^2 + k_2^2)}$$
$$= 4\frac{\sqrt{E_1(E_1 - eV_a)(eV_e - E_1)}}{eV_e(eV_e - eV_a)}\exp(-2\kappa_1 d) \quad . \quad (10.7)$$

The second line is obtained by using (10.3).

10.2.2 Tunneling Matrix Element

Once the value of the transmittance is known, one can calculate the transition matrix element for the tunneling. Consider a situation where the n-type lead is replaced with a second QW, as in the case of a real experimental situation (Fig. 10.1c). The transmittance of the barrier is not affected by this substitution. In order to evaluate the transition matrix element, the escape rate is calculated for an electron from the quantized level in the central QW to a quantized level in the n-side QW. Semi-classically, the escape rate is given by the probability that an electron escapes every time it hits the barrier (transmittance T) multiplied by the frequency f it hits the barrier. The frequency f is given by

$$f = \frac{v}{2W} = \frac{\sqrt{E_1}}{W\sqrt{2m^*}} \quad , \tag{10.8}$$

where $v = dE_1/\hbar dk_1$ is the velocity of the electron in the central QW. Therefore, the electron escape rate is given by

$$R_{\mathrm{esc}} = \frac{4E_1}{W\sqrt{2m^*}} \frac{\sqrt{(E_1 - eV_a)(eV_e - E_1)}}{eV_e(eV_e - eV_a)} \exp[-2\kappa_1 d] \quad . \tag{10.9}$$

Such an escape rate can also be calculated if the tunneling transition matrix element is known. Let $|M_e|^2$ be the matrix element for an electron to make a tunneling transition from a quantized level in the central QW to a state in the n-side QW. For such a resonant tunneling, energy and in-plane momentum of the electron should be conserved:

$$E_1 + \frac{\hbar^2 k_{||}'^2}{2m^*} = eV_a + E_2 + \frac{\hbar^2 k_{||}^2}{2m^*} \quad , \tag{10.10}$$

$$k_{||}' = k_{||} \quad . \tag{10.11}$$

Here, $k_{||}'$, $k_{||}$, and E_2 denote the in-plane momentum of an electron in the central QW state, that in the n-side QW, and the quantized energy level in the n-side QW, respectively.

The transition rate for this process is given by Fermi's golden rule:

$$W_{\mathrm{QW}\rightarrow\mathrm{n}} = \frac{2\pi}{\hbar}|M_e|^2\delta\left(E_2 - E_1 + eV_a\right) \quad , \tag{10.12}$$

when the resonance condition $E_1 = E_2 + eV_a$ is satisfied and zero otherwise. In order to calculate the total escape rate, we need to integrate the transition rate over all momentum states in the n-side QW (denoted by $k_{||}$), and all the momentum states in the central QW (denoted by $k_{||}'$). The total escape rate is given by

$$\begin{aligned} R_{\mathrm{esc}} &= \int \frac{2\pi}{\hbar}|M_e|^2\delta(E_2 - E_1 + eV_a)(2\pi)^2\delta(k_{||} - k_{||}')(2\pi)^2\delta(k_{||}') \\ &\quad \times \frac{1}{(2\pi)^2}d^2k_{||}\frac{1}{(2\pi)^2}d^2k_{||}' \\ &= \frac{2\pi}{\hbar}|M_e|^2\delta(E_2 - E_1 + eV_a) \quad . \end{aligned} \tag{10.13}$$

The δ functions for momentum are inserted to denote parallel momentum conservation and the fact that only $k_{||}' = 0$ states are considered in the central QW, respectively. Equating the right-hand sides of (10.9) and (10.13) yields the explicit expression for the tunneling transition matrix element:

$$|M_e|^2 \delta(E_2 - E_1 + eV_a) =$$
$$\frac{\hbar}{2\pi} \frac{4E_1}{W\sqrt{2m^*}} \frac{\sqrt{E_1 - eV_a}(eV_e - E_1)}{eV_e(eV_e - eV_a)} \exp(-2\kappa_1 d) \tag{10.14}$$

for $E_1 = E_2 + eV_a$ and zero otherwise.

10.2.3 Electron Tunneling Current Density into the Central QW

The n-side QW is populated with electrons, and the electron density determines the quasi-Fermi level, E_{FR}, in this QW. If electrons are populating the central QW, one can assign a quasi-Fermi level for the central QW as well (E_{FL}), as shown in Fig. 10.1c. The quasi-Fermi levels, E_{FL} and E_{FR}, are defined with respect to the bottom of the quantized energy levels, E_1 and E_2, respectively. The electron tunneling current density into the central QW can be calculated by integrating the product of the matrix element and the Fermi function for the entire momentum space of both QWs. Using (10.14) for the expression of the matrix element, the resonant tunneling current density can be evaluated as

$$J_{in} = e \int \frac{2\pi}{\hbar} |M_e|^2 \delta(E_2 - E_1 + eV_a)(2\pi)^2 \delta(\mathbf{k}_{||} - \mathbf{k}'_{||}) f_R(E) \left[1 - f_L(E)\right]$$
$$\times \frac{1}{(2\pi)^2} \mathrm{d}^2\mathbf{k}'_{||} \frac{1}{(2\pi)^2} \mathrm{d}^2\mathbf{k}_{||}$$
$$= \frac{ek_BT}{\pi\hbar^2 W} \frac{\sqrt{2m^*(E_1 - eV_a)}E_1(eV_e - E_1)}{eV_e(eV_e - eV_a)} e^{-2\kappa_1 d}$$
$$\times \frac{e^{-E_{FL}/k_BT}}{e^{-E_{FR}/k_BT} - e^{-E_{FL}/k_BT}} \left(\frac{E_{FL} - E_{FR}}{k_BT} - \ln \frac{1 + e^{-E_{FR}/k_BT}}{1 + e^{-E_{FL}/k_BT}}\right)$$
$$\tag{10.15}$$

when $E_1 = E_2 + eV_a$ and zero otherwise. The resonant tunneling rate can be found from the expression for the current density using the relation $R_T = J_{in}A_{eff}/e$, where A_{eff} is the effective area of the junction.

The expression for resonant tunneling rate calculated here holds only for the tunneling between two ideal QWs without any significant source of linewidth broadening. The tunneling linewidth is infinitely sharp: resonant tunneling is allowed only when the quantized levels of the two QWs align perfectly. Such a strict selection rule results from in-plane momentum conservation. In the low-temperature limit, the resonant tunneling rate is directly proportional to the difference between the two quasi-Fermi levels. In a realistic device, various line-broadening mechanisms exist and the resonant tunneling is never that sharp. In order to estimate the tunneling rate in a real device, one needs the resonant tunneling rate expression in the presence of line-broadening mechanisms.

10.2.4 Effect of Inhomogeneous Broadening

As will be discussed in Chap. 11, the n-side and p-side QWs used in the experiment are alloyed QWs made out of the ternary material AlGaAs. Such an alloyed QW has a significant inhomogeneous linewidth at low temperatures. In order to model this situation, a continuum of QW states was assumed in the n-side QW, the energy distribution of which is given by a simple Gaussian distribution. The probability of finding a QW state in the n-side QW with energy between E_2 and $E_2 + dE$ is given by

$$P(E_2)dE = \frac{1}{\sqrt{2\pi}\sigma} \exp[-(E_2 - E_{20})^2/2\sigma^2]dE \quad , \tag{10.16}$$

where E_{20} and σ denote the mean and the standard deviation of the Gaussian distribution, respectively. It is reasonable to assume that all the states in the n-side QW share a single quasi-Fermi level E_{FR} measured from E_{20}. The total resonant tunneling rate can be found as follows:

- Given the quasi-Fermi levels for the two QWs, use (10.15) to find the resonant tunneling rate of a specific QW state with energy E_2.
- Multiply this by the probability of finding a state between energy E_2 and $E_2 + dE$.
- Integrate this product over the linewidth of the central QW.

There can be several factors that can determine the linewidth of the central QW. The first is the broadening due to electron–hole pair recombination lifetime, which has a time scale on the order of $\sim$10 ns. This usually is masked by the dephasing due to dissipation mechanisms such as acoustic phonon emission. Such dephasing has a time scale on the order of $\sim$10 ps, and corresponds to a homogeneous linewidth of $\sim$10 μeV. The dominant mechanism for the line broadening in the central QW is still inhomogeneous broadening due to interface roughness at the QW boundaries. This should have linewidths of $\sim$0.5 meV, for optimally grown QWs.

Figure 10.2 shows the resonant tunneling rate curves as a function of the external bias voltage, for both electrons (Fig. 10.2a) and light holes (Fig. 10.2b). These curves are calculated using the following parameters: operation temperature $T = 50$ mK, radius of the device $r = 300$ nm, central QW width $W = 10$ nm, barrier thickness for electrons and holes $d = 20$ nm, n-side QW linewidth for electrons $\sigma_e = 2$ meV, p-side QW linewidth for light holes $\sigma_{lh} = 5$ meV, and central QW linewidth $dE = 0.5$ meV. Under these device parameters, the heavy hole tunneling rate is negligible. The resonant tunneling rate curves feature the following shape: since the quasi-Fermi level in the n- and p-side QWs lie in the middle of the broadened line and the operation temperature is low, the rising edge of the tunneling rate curve is roughly linear as a function of the bias voltage, while the falling edge retains the Gaussian shape of the broadened distribution.

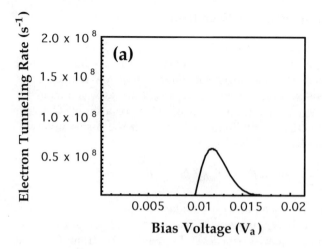

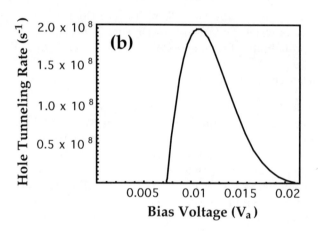

Fig. 10.2 a, b. Resonant tunneling rate curves as a function of the external bias voltage, calculated by the method described in this section. (**a**) Electron resonant tunneling rate curve. (**b**) Light hole resonant tunneling rate curve

10.3 Coulomb Blockade Effect on Resonant Tunneling

The previous section discussed in detail how one can calculate the resonant tunneling rates of electrons and holes as a function of the external bias voltage. The electron and hole resonant tunneling rates can be uniquely determined if the quasi-Fermi level of the central QW, as well as those for the n- and p-side QWs, are given. The quasi-Fermi levels for the n- and p-side QWs can be assumed to have fixed values even after electron/hole tunneling events, since the carriers are supplied from highly doped layers nearby. The quasi-Fermi level for the central QW depends on the number of carriers in the QW. In a real turnstile device, the wafers are designed so that the first electron tunneling occurs at a lower bias voltage compared to the first hole tunneling event. Therefore, at bias voltages of interest, it is assumed that the central QW is always populated with a finite number of electrons.

As more electrons tunnel into the central QW, the quasi-Fermi level in the central QW increases due to the Pauli exclusion principle: since only one electron can occupy a single quantum state, the electrons that are supplied later necessarily occupy quantum states with higher energies. For a p–n junction with such a small junction area as that considered here, there is another reason that causes the effective quasi-Fermi level in the central QW to increase. The capacitance between the central QW and the side QWs is very small in these small junctions, and the charging energy associated with such a small capacitance can be significant. As an extra electron tunnels into the central QW, the capacitor between the central QW and the n-side QW becomes discharged by e. Since the total voltage bias across the p–n junction is fixed by the external constant voltage source, such change in the charge configuration results in the rearrangement of the voltage drop across the tunnel barriers. Simple analysis shows that when an extra electron tunnels into the central QW, the voltage drop across the n-side barrier decreases by $e/(C_n + C_p)$, while that across the p-side barrier increases by the same amount.

This voltage rearrangement can simply be viewed as all of the energy levels of the central QW shifting upwards by $e^2/(C_n + C_p)$. In terms of the resonant tunneling rates, this means that the electron resonant tunneling rate curve is shifted upwards by $e/(C_n + C_p)$ in applied bias voltage, while the hole resonant tunneling rate curve is shifted downwards by the same amount when an extra electron tunnels. Such a shift can be understood easily by the Coulomb interaction: since the central QW now carries more negative charge, the Coulomb repulsive interaction makes it more difficult for electrons to tunnel, while the Coulomb attractive interaction makes it easier for the holes to tunnel. This shift is only sensitive to the total charge in the central QW. When a hole tunnels, it neutralizes one electron and the curves shift in the opposite directions.

Sections 10.1 and 10.2 described how to calculate the resonant tunneling rates for electrons and holes in the structure under consideration from microscopic principles, and discussed the effect of charging energy. Based on the arguments given in these two sections, the current transport in such a structure is considered under various bias voltage conditions. The behavior of the junction under constant voltage bias is simplified, and the following rules are derived, which will be used to analyze the current transport and photon emission:

- Electron and hole resonant tunneling rate curves as a function of bias voltage have the shape shown in Fig. 10.2. The rising edge increases linearly as a function of bias voltage, while the falling edge retains a Gaussian lineshape due to inhomogeneous broadening.

- As the bias voltage is increased, the electron resonant tunneling condition is satisfied first. Therefore, the central QW is always populated with electrons.
- When an electron tunnels, the electron resonant tunneling rate curve shifts upwards in applied bias voltage by $e/(C_n + C_p)$, while the hole resonant tunneling rate curve shifts downwards by the same amount.
- When a hole tunnels, the resonant tunneling rate curves shift in the opposite direction.
- The shift of the resonant tunneling rate curves are sensitive only to the total charge in the central QW. Photon emission does not change the charge state, so there is no shift in the resonant tunneling rate curves associated with photon emission.

10.4 Coulomb Staircase

10.4.1 DC Voltage Bias Condition

First, consider the DC operation of a double-barrier p–i–n junction under constant voltage bias condition. In the ideal case where nonradiative recombination is absent, current flows through the device only by radiative recombination of electron–hole pairs in the central QW. This is possible only if both carrier (electrons and holes) tunneling events are present. At a given DC bias voltage, the electrons charge up the central QW (and move the electron tunneling rate curves up and hole tunneling rate curves down) until no more electron tunneling is allowed. If, at this voltage, the hole tunneling rate is zero, current does not flow. One needs to increase the DC voltage further to approach the hole tunneling rate curve until the electron and hole tunneling rate curves meet at the DC bias voltage, as shown in Fig. 10.3. In this figure, the rising edges of the electron and hole resonant tunneling rate curves are exaggerated, and the hole resonant tunneling rate curve has a larger slope.

When the DC bias voltage is lower than point A, the hole tunneling is not allowed with $m - 1$ electrons in the central QW (where the electron and hole tunneling rates are given by the solid lines in Fig. 10.3). Once the DC bias voltage exceeds point A (with $m - 1$ electrons in the central QW), the mth electron tunneling is allowed with a finite probability and the tunneling rate curves shift to the broken lines (m electrons in the central QW). As soon as this electron tunnels, further electron tunneling is completely suppressed, while the hole tunneling rate becomes very high. A hole tunnels immediately, and the resonant tunneling rate curves return to the solid lines. In this case, electron tunneling is the slowest process that initiates the current flow, and hole tunneling follows immediately. In the limit that the hole tunneling rate is very large, so that the the average waiting time for hole tunneling is completely negligible, electron injection becomes a random Poisson point process

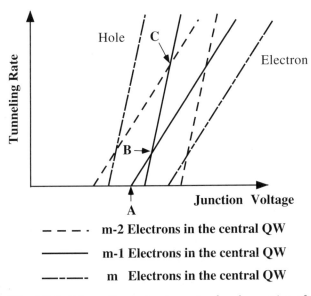

Fig. 10.3. Schematic carrier dynamics for observation of the Coulomb staircase in DC I–V characteristics. The *broken lines, solid lines* and *dashed lines* are electron and hole resonant tunneling rates when m, $m - 1$ and $m - 2$ electrons are present in the central quantum well (QW), respectively

and photon emission becomes completely random. The average DC current is proportional to the electron tunneling rate, given by

$$I(V) = e\,R_e(V, m - 1) \quad, \tag{10.17}$$

where $R_e(V, m - 1)$ denotes the electron tunneling rate at voltage V when the central QW is populated with $m - 1$ electrons.

As the bias voltage is increased above point B, both electron and hole can tunnel, but the hole tunneling rate is now larger than the electron tunneling rate. Occasionally an (the mth) electron tunneling occurs first, in which case the current flows by the same mechanism as between points A and B discussed above. Most of the time a (the first) hole tunnels first, and the electron and hole resonant tunneling rate curves shift to the dashed lines (with $m - 2$ electrons in the central QW). Further hole tunneling is completely suppressed, and the electron tunneling rate becomes higher. An electron tunneling will follow, and the tunneling rate curves are shifted back to the solid lines. Since the hole tunneling rate (given by the solid line) is smaller than the electron tunneling rate (given by the dashed line) between points B and C, the hole tunneling triggers the current flow in this case, and electron tunneling follows immediately. The average DC current in this section is proportional to the hole tunneling rate, and is given by

$$I(V) = e\,R_h(V, m - 2) \quad, \tag{10.18}$$

where $R_{\rm h}(V, m-2)$ denotes the hole tunneling rate at voltage V when there are $m-2$ electrons in the central QW. Since this curve has a larger slope compared to the hole resonant tunneling rate curve, the I–V curve features a steep linear increase. Since hole tunneling is a Poisson point process, photon emission is not regulated.

When the bias voltage is increased above point C, a (the first) hole tunneling event (rate given by the solid line) occurs much faster than an [the $(m-2)$th] electron can tunnel (rate given by the dashed line), and the device is mostly waiting for an electron to tunnel and shift the resonant tunneling curves back to the solid lines. Once the electron tunnels, a hole tunnels immediately afterwards. The current is triggered by an electron again and features a slow linear increase following the dashed electron tunneling rate curve.

Therefore, the I–V characteristics of such a device under DC bias voltage condition feature staircase-like behavior, alternating between a slow slope for the electron tunneling rate curve and a steep slope for the hole tunneling rate curve. The steps reflect the fact that the number of electrons in the central QW decreases by 1 and appear every time the bias voltage is increased by $e/C_{\rm p}$. The height of the steps (in current) is determined by the slope of the resonant tunneling rate curve. This is very similar to the situation in an asymmetric M–I–M–I–M double tunnel junction, where the Coulomb staircase was first observed [12, 66]. It should be noted that the Coulomb staircase is expected when the tunneling rate for the two tunnel junctions are different for the M–I–M–I–M junction case. In this case, the requirement is that slopes on the rising edges of the resonant tunneling rate curves are different. Although the example presented in this section treats the case where the hole resonant tunneling rate curves have a steeper slope, a similar argument applies to the case where the electron tunneling rate has a steeper slope.

10.4.2 DC + AC Voltage Bias Condition

Next we consider the case where an AC square-wave voltage signal with amplitude ΔV and frequency $f = 1/T$ (T is the period) is added on top of the DC bias voltage. Unlike the DC voltage bias case, since the junction voltage is modulated between V and $V + \Delta V$, the current starts to flow at a voltage before the electron and hole tunneling rate curves start to overlap. The situation is shown schematically in Fig. 10.4, where electron tunneling occurs at the voltage V and hole tunneling occurs at voltage $V + \Delta V$. We further assume that the hole resonant tunneling rate curve has a sharp rising edge, so that the hole tunneling rate multiplied by the half period is always much larger than one ($R_{\rm h}(V, k)T/2 \gg 1$, where k is the number of electrons in the central QW). This guarantees that the net electron number in the central QW is reset to the initial value during the on-pulse voltage, $V + \Delta V$, every modulation cycle, and the junction operation is purely determined by the electron tunnelling conditions. Under such assumption, there are two regimes

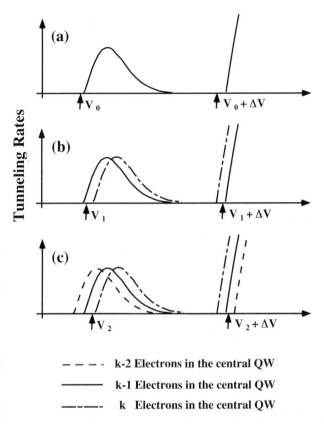

Tunneling Rates

(a)

V_0 $V_0 + \Delta V$

(b)

V_1 $V_1 + \Delta V$

(c)

V_2 $V_2 + \Delta V$

— — — · **k-2 Electrons in the central QW**

————— **k-1 Electrons in the central QW**

—·—·— **k Electrons in the central QW**

Fig. 10.4a–c. Schematic carrier dynamics for observation of the Coulomb staircase in DC + AC I–V characteristics. The *broken lines, solid lines* and *dashed lines* are electron and hole resonant tunneling rates when k, $k - 1$ and $k - 2$ electrons are present in the central quantum well (QW), respectively

of operation, depending on the frequency of the AC modulation voltage signal. In this section, the high-frequency limit is considered, where frequency is high (the period is short) so that the tunneling probability of an electron during the off-pulse duration is much smaller than unity $(R_e(V, k)T/2 \ll 1)$.

Figure 10.4(a) shows the situation where the junction voltage is modulated between V_0 and $V_0 + \Delta V$. When the voltage is at V_0, the central QW is filled with $k - 1$ electrons, and further electron tunneling is suppressed. At $V_0 + \Delta V$ the hole tunneling rate curve is not reached and no holes are allowed to tunnel; thus, one sees no current flowing through the device.

When the DC bias voltage is slightly increased to V_1, an (the kth) electron is now allowed to tunnel at V_1 (Fig. 10.4b). However, the operation frequency is high, and the electron tunneling probability during the off-pulse duration satisfies $R_e(V_1, k - 1)T/2 \ll 1$. Whether the electron tunnels or not, the tunneling rate curves are set to an identical condition (solid lines) each time the junction voltage is switched to $V_1 + \Delta V$ since the (first) hole tunneling probability at the on-pulse is close to unity when the electron tunnels (broken lines for k electrons in the central QW) and is zero when the electron does not tunnel (solid lines with $k - 1$ electrons). Therefore, the current is determined by the fraction of periods where an electron tunnels multiplied by the frequency of the modulation input, given by

$$I = eR_{\mathrm{e}}(V_1, k - 1)/2 \quad . \tag{10.19}$$

The factor $1/2$ simply reflects the fact that electron tunneling is allowed for only half of the modulation period. In this case, the current increases linearly with increasing DC bias voltage and is independent of the AC frequency.

When the DC bias voltage is further increased to V_2 and exceeds the rising edge of the hole tunneling rate curve with $k - 1$ electrons in the central QW, the first hole is allowed to tunnel during the on-pulse (at $V = V_2 + \Delta V$) with close to unity probability (Fig. 10.4c). Then, the resonant tunneling rate curves for both electrons and holes shift to the dashed lines (with $k - 2$ electrons in the central QW), and when the junction voltage is decreased to V_2, the electron tunneling rate is now given by the dashed line, which is higher than the solid line. Since the operation frequency is high, the $(k-1)$th and $(k - 2)$th electron tunneling probabilities satisfy $R_{\mathrm{e}}(V_2, k - 1)T/2 \ll R_{\mathrm{e}}(V_2, k - 2)T/2 \ll 1$, and most of the time only one electron tunnels. Since $R_{\mathrm{e}}(V_2, k - 2) > R_{\mathrm{h}}(V_2 + \Delta V, k - 1)$ the current is now given by

$$I = eR_{\mathrm{h}}(V_2 + \Delta V, k - 2)/2 \quad . \tag{10.20}$$

The transition between Fig. 10.4b and c occurs as the voltage for the on-pulse crosses a sharp rising edge of the hole resonant tunneling rate curve and the I–V characteristics feature a step-like increase.

A similar increase in the electron tunneling rate is expected whenever another electron is compensated by the addition of a hole to the central QW, due to the shift in the tunneling rate curves. Just like in the DC voltage bias case, the steps in the I–V characteristics occur at voltage values where an additional hole is added to the central QW and the corresponding current values are determined by the change in electron tunneling rate when a hole is added to the central QW. Therefore, the value of the current step is independent of the modulation frequency. Also, since the electron/hole injection is close to a Poisson point process, photon emission is also not controlled in time.

10.5 Turnstile Operation

In this section, we consider the low-frequency limit where the frequency of the AC square modulation is low (the period is short) so that the electron tunneling rate integrated over the off-pulse duration is much larger than unity

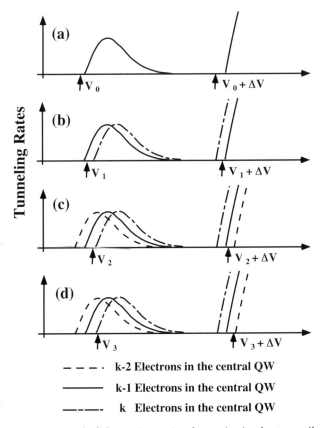

Fig. 10.5a–d. Schematic carrier dynamics in the turnstile operation regime when the junction is biased with a DC + AC voltage source in the low-frequency limit. The *broken lines*, *solid lines* and *dashed lines* are electron and hole resonant tunneling rates when k, $k-1$ and $k-2$ electrons are present in the central quantum well (QW), respectively

$(R_\mathrm{e}(V,k)T/2 \gg 1$, where k is the number of electrons in the central QW). This means that at least one electron is guaranteed to tunnel into the central QW during the off-pulse. When enough additional electrons are added to the central QW by tunneling, the shift of the tunneling rate curve due to Coulomb blockade effect causes the tunneling rate to change dramatically, so that the tunneling probability is completely suppressed, i.e., $R_\mathrm{e}(V,k-n)T/2 \ll 1$ where n is the number of electrons that are required to tunnel before further electron tunneling is completely suppressed. This is a regime where the number of carriers injected per modulation period is well defined due to the Coulomb blockade effect.

Figure 10.5a shows the case where the junction voltage is modulated between V_0 and $V_0 + \Delta V$. At V_0, $k-1$ electrons are present in the central QW, and the tunneling rate curves are given by the solid lines. When the junction voltage is increased to $V_0 + \Delta V$ by the AC modulation voltage, it is not high enough to hit the hole resonant tunneling rate curve for $k - 1$ electrons in the central QW, and no hole tunneling occurs. Therefore, no current flows through the device.

When the DC voltage is increased slightly, as shown in Fig. 10.5b and Fig. 10.6a, the kth electron tunnels at a bias voltage of V_1. The schematic band diagram under this bias condition is shown in Fig. 10.6a. Since the modulation frequency is low and the electron tunneling rate integrated over the off-pulse duration is large, the kth electron tunnels into the central QW. The resonant tunneling rate curves for electrons and holes shift to the broken lines for k electrons in the central QW. Once this electron tunnels, the tunneling rate for the next electron is reduced close to zero, and further electron tunneling is suppressed. When the junction voltage is increased to $V_1 + \Delta V$, hole tunneling is allowed, as shown schematically in Fig. 10.6b. Just like in the previous section, it is assumed that the hole tunneling rate at $V_1 + \Delta V$ is large enough so that the probability of the first hole tunneling during the on-pulse duration is close to unity. A single hole tunneling event neutralizes one electron in the central QW and shifts the tunneling rate curves back to the solid lines ($k - 1$ electrons in the central QW). Further hole tunneling is suppressed during the on-pulse duration. This first hole recombines with an electron in the central QW and emits a single photon.

When the DC junction voltage is further increased (so that $V_2 + \Delta V$ allows a hole to tunnel with $k-1$ electrons in the central QW) as in Fig. 10.5c, two holes are allowed to tunnel during the duration of the on-pulse ($V = V_2 + \Delta V$) with close to unity probability. After the two holes tunnel, the tunneling rate curves move to the dashed lines ($k - 2$ electrons in the central QW), and further hole tunneling is suppressed. When the junction voltage is decreased to V_2, two electrons are allowed to tunnel (with probability close to unity, since the second electron tunneling rate is also large) before further electron tunneling is suppressed. After two electrons tunnel, the charge state of central QW is returned to the initial condition. This will result in the emission of two photons per modulation cycle.

A similar argument can be applied to the three-hole tunneling case. A schematic of this operation is given in Fig. 10.5d. We note that when the frequency of the AC modulation signal is low the number of electrons and holes injected per modulation period is very well defined, since the electron and hole tunneling probabilities are either close to unity or completely suppressed. This gives rise to well-defined plateaus in the I–V curve, with each plateau corresponding to $I = nef$. The transition between two adjacent plateaus is rather sharp, since it is determined by the slope of the rising edge in the electron and hole tunneling rate curves. This is called turnstile operation, since the number of electrons and holes injected per modulation period, and thus the number of photons emitted per modulation period, is a well-

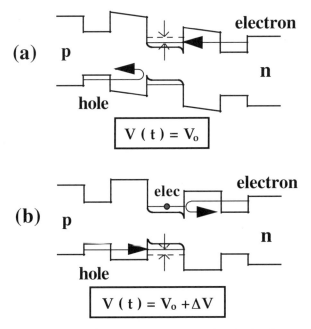

Fig. 10.6a, b. Schematic operation of a single-photon turnstile device, with one electron and one hole injected per modulation period

defined number. The fluctuation of photon number per modulation period is suppressed by the Coulomb blockade effect.

A mesoscopic p–n junction that operates in this regime with $n = 1$ is called a single-photon turnstile device and was first proposed by Imamoḡlu and Yamamoto in 1994 (Fig. 10.6) [104]. In the original proposal, the resonant tunneling linewidth for both electrons and holes were assumed to be narrow compared to the single-electron (-hole) charging energy of $e^2/2C_i$ $(i = n, p)$. This means that whenever a resonant tunneling condition is satisfied for an electron (hole) a single-electron (-hole) tunneling event will guarantee that the next electron (hole) tunneling rate is completely suppressed. This requirement that the resonant tunneling linewidth be narrower than the Coulomb charging energy is not easy to meet in a realistic device, due to various linewidth broadening mechanisms. It is very important from an experimental point of view that such a requirement be removed in order to realize a single-photon turnstile device. It turns out that the assumption of narrow tunneling linewidth is not a fundamental requirement. In Sect. 10.2, an inhomogeneously broadened tunneling linewidth is introduced that is much broader than the single charging energy when calculating the resonant tunneling rates. The device operation principles discussed in this section show how single-photon turnstile operation is still possible with resonant tunneling linewidths much larger than the single electron charging energy, as long as the temperature is low enough to satisfy $e^2/C_i \gg k_\mathrm{B}T$. It is worth noting that such a broad resonant tunneling linewidth also opens up the possibility for multiple- photon generation per modulation period, which was not predicted by the original proposal of the single-photon turnstile device.

10.6 Monte-Carlo Simulations

Monte-Carlo numerical simulations [101] were performed to confirm the theoretical model and to show the transition from Coulomb staircase operation to turnstile operation [21]. The details of the simulation will not be reproduced here. The simulation results quantitatively verify the physics of current transport in the mesoscopic p–i–n junction discussed in this chapter and agree with the experimental observations described in Chap. 11.

10.7 Summary

This chapter discussed in detail the theoretical analysis of charge transport and photon emission characteristics of a mesoscopic p–i–n junction shown in Fig. 10.1a. A simplified model was presented to calculate the resonant tunneling rates of electrons and holes as a function of the applied bias voltage and derive a set of rules that govern the electron and hole tunneling events. Based on these rules, the carrier transport properties of these junctions were discussed, under a simple DC bias voltage and when an AC square-modulation voltage is superimposed on top. The analysis predicts observation of the Coulomb staircase in the current–voltage characteristics for the DC bias and the high-frequency limit of the AC bias and a single-photon turnstile operation for the low-frequency limit of the AC bias. Chapter 11 will describe the experimental effort towards the observation of Coulomb blockade effect and realization of the single-photon turnstile device.

11. Single-Photon Generation in a Single-Photon Turnstile Device

11.1 Introduction

The model and discussion presented in Chap. 10 demonstrate the possibility of observing the Coulomb blockade effect in mesoscopic p–n junctions and further imply the possible realization of a "single-photon turnstile device" that generates single photons with a well-defined time interval. Since the arrival time of each photon can be predicted, such single photon states are called "heralded single photons" [104, 269]. This chapter describes the experimental progress towards realization of such single-photon states using a mesoscopic semiconductor p–n junction. The first half of the chapter discusses the fabrication of a mesoscopic p–n junction in a GaAs material system. The second half describes the electrical and optical characteristics of these devices measured at low temperatures. A Coulomb staircase was observed in the electrical measurement, which clearly demonstrates the Coulomb blockade effect present in these devices. Device operation in the turnstile mode indicates that the time interval between single photons is regulated to better than the Poissonian limit. This implies that nonclassical single photons can be generated from a mesoscopic p–n junction utilizing the Coulomb blockade effect.

11.2 Device Fabrication

Fabrication of sub-micron size p–n junctions in the GaAs system that could be used as a light emitter is a challenging proposition for several reasons. The general strategy would be to first grow a vertical p–n junction structure, and then pattern small posts using microfabrication technologies. The pattern can be generated using standard electron-beam lithography technology, but directional (anisotropic) etching with low damage and high spatial resolution is the first challenge. A second challenge is to make an electrical contact to the top of these individual, sub-micron posts and make the contact transparent to optical photons. The third and most difficult challenge is to ensure that the internal quantum efficiency is high enough for an electron–hole pair to produce a single photon in such a small device. This is difficult because it is

Table 11.1. Detailed structure of the double-barrier p–i–n junction grown for the single-photon turnstile device.

Type	Material	Role of Layer	Al Comp.	Doping density	Thickness(Å)
p+	GaAs	Cap layer	0	5×10^{19} cm^{-3} (max. doping)	100
p+	Al$_x$Ga$_{1-x}$As	Contact layer	$x = 0.16$	5×10^{19} cm^{-3} (max. doping)	4000
p	Al$_x$Ga$_{1-x}$As	p-lead	$x = 0.16$	5×10^{18} cm^{-3}	2000
i	Al$_y$Ga$_{1-y}$As	p-side QW	$y = 0.12$	0	75
i	Al$_z$Ga$_{1-z}$As	Barrier	$z = 0.18$	0	200
i	GaAs	Central QW	0	0	100
i	Al$_z$Ga$_{1-z}$As	Barrier	$z = 0.18$	0	200
i	Al$_y$Ga$_{1-y}$As	n-side QW	$y = 0.12$	0	90
n	Al$_x$Ga$_{1-x}$As	n-lead	$x = 0.16$	2×10^{17} cm^{-3}	10000
n+	GaAs	Buffer layer	0	5×10^{18} cm^{-3}	5000
n+	GaAs	Substrate	0	5×10^{18} cm^{-3}	

known that the nonradiative recombination of an electron–hole pair through surface states in GaAs becomes a dominant mechanism in the radiative recombination as the device's size shrinks [155]. This section describes how a mesoscopic p–n junction was fabricated, and how these challenges were met.

11.2.1 Wafer Design and Growth

The resonant tunneling rates of electrons and holes for the structure shown in Fig. 10.1a can be calculated as discussed in Chap. 10. The thicknesses of the three quantum wells (QWs) and the two tunnel barriers need to be determined carefully, so that the electron and hole resonant tunneling conditions are well-separated in bias voltage and their respective tunneling rates are reasonably large. Table 11.1 shows the layers of the double-barrier p–i–n junction grown for the experiment. The wafer was grown using the molecular beam epitaxy (MBE) technique by our colleagues Dr. Kan and Mr. Watanabe, at Hamamatsu Photonics, Japan. We employed a digital alloying technique to form Al$_x$Ga$_{1-x}$As alloys for the tunnel barriers and alloy QWs for the p-side and n-side. This is where layers of AlAs and GaAs are alternately grown to form AlGaAs layers where the relative thickness of the layers determines the Al composition [219]. Such a digital alloying technique is known to provide better barriers than a random alloy, since random scattering in the alloy is replaced with more regulated interface scattering. It is known that, provided that the period of digital alloying is small enough, the photoluminescence linewidth of a QW sandwiched between digitally alloyed barriers features sharper linewidths.

For the structure above, one needs to estimate the bias voltage values where the electron and hole tunneling conditions are satisfied. This could be done by solving the Poisson equation, given the doping concentration of each layer and the overall applied bias voltage. To be strict, one needs to find a self-consistent solution using both the Poisson equation and the Schrödinger equation, since the Poisson equation solves the voltage profile given the charge distribution, while the Schrödinger equation solves the electron wavefunction (which gives the charge distribution) given the voltage profile. Such a self-consistent solution is usually found using the mean field approximation (also known as the Hartree approximation). Rather than trying to solve these equations numerically, a very simple model was used here to estimate the resonant tunneling conditions. In the current structure, the carriers are supplied to the intrinsic region only by tunneling, and the number of carriers in this part of the junction is very small. The effect of electron tunneling into the central QW on the junction voltage profile is modeled by the charging energy. When an extra electron is in the central QW, the voltage drop across the tunnel barriers is modified as discussed in Chap. 10. Use of such a simple model is justified if electrons (or holes) are confined to the central QW in the intrinsic region and the number of charges that can be stored in the central QW is relatively small.

From the solution of the Poisson equation in the absence of any charge in the intrinsic region, one can calculate the bias voltage at which the first electron is allowed to resonantly tunnel into the central QW. The calculation is done assuming low temperature (which determines the bandgap of GaAs, and thus the built-in potential of the junction, and makes the Fermi function a step function) and yields the value $V_{ext} = 1.526$ V. Similarly, a first hole resonant tunneling is allowed at $V_{ext} = 1.589$ V, if no electrons are in the central QW. The difference between these resonant tunneling voltages is ~63 meV.

11.2.2 Ohmic Contact Formation

The first step in the fabrication process is to form an n-type contact on the substrate side of the wafer. the n-type contact is formed by evaporating an appropriate alloy of Ni, Au and Ge [251]. A metal–semiconductor junction always forms a Schottky barrier in GaAs, since the Fermi level is pinned near the middle of the gap in GaAs. The carrier transport through a Schottky barrier is not Ohmic, i.e., the current–voltage characteristics are not linear, since the carriers need to be thermally injected above the Schottky barrier. The width of the Schottky barrier is determined by the doping density, N, of either donors or acceptors and is $\sim 1/\sqrt{N}$. The only way to make a reasonable Ohmic contact is to increase the doping density in the semiconductor near the contact and make the Schottky barrier thinner. When the barrier becomes thin enough, the electron (or hole) transport will be dominated by charge tunneling through the Schottky barrier, and the current–voltage characteristics become linear. For p-type contacts, one can usually dope p-type layers high enough ($\sim 10^{19}$ cm^{-3}) in the growth procedure so that this is not an issue. For n-type layers, doping levels are limited to $\sim 10^{18}$ cm^{-3}, and good Ohmic contacts cannot be formed. One way to achieve high n-doping is to evaporate Au and Ge on GaAs and heat the sample up to $\sim 450^\circ$C. Ga atoms then diffuse into the Au, creating empty Ga sites, which become available for the Ge atoms to fill. Through such substitution, a very high n-type doping density can be achieved near the surface. Ni helps Ge atoms to diffuse into the GaAs crystal better, as well as playing the role of a wetting agent that prevents "balling up" of Au/Ge metal contacts. The temperature and duration of the anneal is a critical factor in determining the contact resistance. The anneal was done at 450°C for 30 s in a rapid thermal annealer (RTA).

11.2.3 Device Definition: Electron-Beam Lithography

The electron-beam lithography technique was used to define a small junction area. Electron-beam lithography is a technique where a resist sensitive to electron-beam exposure is put down on the sample, and patterns are drawn by scanning a focused electron beam on the resist. Since a high-energy (30–40 keV) electron beam can have a significantly shorter wavelength (~ 0.007 nm) compared to ultraviolet photons used in conventional optical lithography (~ 250 nm), the diffraction-limited minimum feature size that can be defined with electron-beam lithography is much smaller than with optical lithography. The resolution of electron-beam lithography is practically limited by resist exposure due to secondary electrons that are emitted when a high-energy electron bombards the sample. There are several types of electron-beam resists available, both positive and negative. Polymethylmethacrylate (PMMA), a well-known positive resist with very high resolution, was used. PMMA is a large polymer molecule, and some bonds can be broken with electron-beam exposure. When the bonds are broken, the resulting smaller molecules can be dissolved by certain solvents (called developers). There are PMMAs with different molecular weights that have different development rates. Two types of PMMA were used, with molecular weights of 495 K and 960 K. For the developer, a 3:1 mixture of isopropanol and methylisobutylketone (MIBK) is commonly used. For electron-beam writing, a scanning electron microscope (Leica StereoScan 440) converted into an electron-beam writing system using external pattern generation and a scan control system (Nabity Systems) was used. The smallest linewidth we could achieve with this system was about 40 nm. Circular devices with radii between 200 and 1000 nm were defined using this process.

11.2.4 Metal Evaporation and Liftoff

The electron-beam lithography is performed in a way that is optimized for a metal liftoff technique. Figure 11.1 shows the liftoff procedure using electron-beam lithography. First, one spins on a layer of resist and patterns the resist

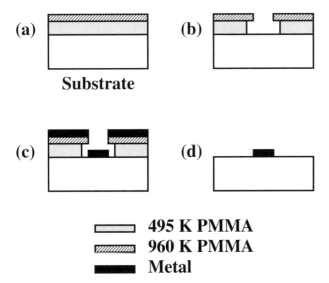

Fig. 11.1 a–d. Metal liftoff process using bilayer PMMA. (a) Bilayer PMMA film on a GaAs sample. (b) After electron-beam lithography, desired parts of the PMMA film have been removed. The use of bilayer PMMA creates an undercut profile in the resist. (c) Evaporation of metal. Line-of-sight deposition leaves patterns of metal on the wafer it are disconnected from the film on top of the resist. (d) After soaking in solvent, the resist and the metal film on top of that is removed and only the desired pattern of metal remains on the sample

to open up holes where one wants the metal to be put down. Then, a thin layer of metal is evaporated over the whole sample. Evaporation is a way to deposit thin films in a nonconformal way, i.e., metal atoms move in a straight path until they hit a surface where they stick. Therefore, the vertical side walls are not covered with metal. Then, the sample is soaked in a solvent that dissolves the resist. The metal sitting on top of the resist is removed from the sample, while the portion directly deposited on top of the sample remains. This way, one can leave a desired pattern of metal on the sample surface.

In order to enhance the liftoff process, it is very important that the metal film on top of the resist is not connected to the metal on the sample surface. Therefore, it is desirable to have a resist profile that features an undercut (overhang) structure [248]. In electron beam lithography, such a resist profile is naturally formed due to resist exposure by secondary electrons (called "proximity effect") [249]. Such undercut features can easily be enhanced further using a bilayer of PMMA. In this process, PMMA with a smaller molecular weight (e.g., 495 K) is put down on the bottom and PMMA with larger molecular weight (e.g., 960 K) is put down on top. It should be noted that a typical solvent for PMMA is chlorobenzene; however, this is such a good solvent that an attempt to put down the second layer will completely dissolve the first layer. Therefore, 960 K PMMA in MIBK is used for the second layer. After exposure, the 495 K PMMA develops faster than the 960 K PMMA, and an undercut resist profile can easily be achieved. In the fabrication process discussed here, 100 Å of Cr was evaporated, followed by 300 Å of Au. This metal serves as a p-type contact, since the GaAs layer near the surface of the wafer is very highly doped.

11.2.5 Device Isolation: ECR-RIE

The metal circles left on the sample define the junctions. The next step is to isolate the devices from one another. This is done by a self-aligned etching process using a plasma. It is called "self-aligned" because no independent step is used to define the top contact: in this process, the Cr/Au metal circles serve both as the mask for etching and as the p-type ohmic contact. Since the p–i–n junction is located about 6000 Å below the surface (Table 11.1), the etch depth had to be at least 7000 Å. For the smallest devices defined, this would mean that we need etching techniques with an aspect ratio better than 5:1. Wet chemical etching cannot provide such high aspect-ratio etching; one has to use dry etching. The etching was done by reactive ion etching (RIE) in a high density of Cl_2/BCl_3 plasma generated by an electron cyclotron resonance (ECR) source. RIE is a dry-etching technique in which an etching ion (in this case Cl) chemically reacts with the GaAs crystal and forms a volatile species that can be pumped away in a vacuum chamber.

Conventional RIE utilizes the plasma sheath that develops between two parallel plates (or between a chuck that holds the sample and the remainder of the chamber) when a radio-frequency (RF) bias (at 13.56 MHz) is applied between the two plates [250]. In such RIE, the density of the plasma is determined by the applied RF power. At the same time, a DC voltage develops between the plasma sheath and the sample chuck (which is always used as one of the parallel plates) that accelerates the reactive ions towards the sample, and directional (anisotropic) etching can be performed. However, this DC voltage is also determined by the RF power and is on the order of 300 V. Such a high acceleration voltage induces serious damage in the sample due to ion bombardment, but is unavoidable since a certain RF power is required to generate any plasma at all.

ECR is one of many popular technologies to generate a very-high-density plasma. When a microwave source with a frequency of 2.45 GHz is applied to a gas in a magnetic field gradient, the cyclotron motion of the electrons (with angular frequency given by $\omega_c = eB/m$, where B is the magnetic field and m is the electron mass) becomes resonant with this microwave frequency when the magnetic field is 875 gauss. This means that at the plane where the magnetic field is 875 gauss (called the "resonant zone") the electrons absorb the microwave power resonantly and generate a very-high-density plasma. The pressure should be low enough (1 ∼ 20 mtorr) so that the collision

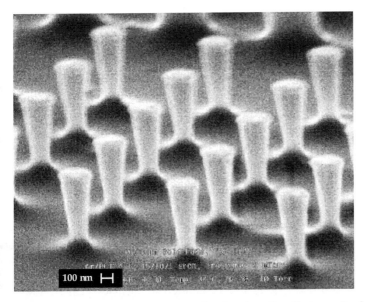

Fig. 11.2. Typical post structure of p–i–n junction that can be fabricated using the fabrication technology described in this section

rate of electrons with ions is small compared to the cyclotron frequency. The generated density of plasma is determined by the microwave power. Unlike conventional RIE, an ECR plasma does not generate a high bias voltage toward the sample chuck. An independent RF power source can be applied to the chuck, which controls the DC acceleration voltage for the ions to hit the sample. Since the RF power is not used to generate the plasma, the RF power can be decreased arbitrarily to decrease the acceleration voltage. In an ECR plasma, there are two independent parameters to control the plasma density (microwave power) and the DC bias to the sample (RF power).

In the fabrication process used for the mesoscopic p–i–n junctions, the following parameters were used: 15 sccm (unit of gas flow: standard cubic centimeters per minute) of Ar, 10 sccm of BCl_3, and 1 sccm of Cl_2 were supplied to the chamber, and the pressure was maintained at 2 mtorr. 400 W of microwave power was used, and 35 W of RF power was used to generate 45 V of DC voltage. The etch rate was about 1200 Å/min. Figure 11.2 shows typical post structures that can be fabricated using the electron-beam lithography and ECR-RIE technology.

11.2.6 Surface Passivation

One major problem with the idea of a small (sub-micron size) p–n junction light-emitting device made of GaAs is the high nonradiative surface recombination. It is well-known that there is a high density of surface states in GaAs very close to the middle of the energy gap (0.8 eV below the bottom of the conduction band) [155, 221]. Such states not only pin the Fermi energy at the middle of the gap at the semiconductor–air interface, but also act as nonradiative recombination centers for electron–hole pairs. The presence of surface states can be overcome in certain applications involving only one type of carrier (unipolar devices), but always puts a limitation on the performance of devices where both type of carriers are necessary (bipolar devices). For instance, in a metal–semiconductor field- effect transistor (MESFET), the metal gate deposited on the semiconductor surface forms a Schottky contact. The high density of electrons from the metal always saturates the surface states. Since there is no charging/discharging of the surface states on a short time scale, these states simply act as a surface charge density and do not affect the transistor performance at high frequencies [92]. At lower frequencies, charging and discharging of such states is known to produce the well-known $1/f$ noise. This is the reason why metal–insulator–semiconductor field effect transistors do not perform well with GaAs, since the surface states are not saturated. On the other hand, leakage current mediated by the surface states between the collector and the base is the limiting factor for the performance of heterojunction barrier transistors (HBTs) made of n–p–n (or p–n–p) GaAs/AlGaAs heterostructures [167, 201].

Nonradiative recombination of electron–hole pairs through surface states is the major obstacle in achieving high-performance light-emitting devices. The threshold of many edge-emitting and surface-emitting semiconductor lasers are limited by surface leakage current. Low threshold in transverse-junction- stripe (TJS) lasers is achieved by their unique design, where exposure of the p–n junction (and thus the active region) to surface is minimal [180]. Overcoming surface recombination has been the major theme in lowering the threshold of vertical-cavity surface-emitting lasers (VCSELs); the best performance so far has been achieved by confining the injection current to the center of the mesa structure by means of an oxidized AlAs layer [52, 61, 207]. Microlasers using whispering-gallery modes of a semiconductor disc also suffer from surface recombination [96]. The surface recombination is characterized by a surface recombination velocity, S_v (in cm/s), defined from the boundary condition (at $x = 0$) [222]

$$eD_a \frac{\partial n_a}{\partial x}\bigg|_{x=0} = eS_v[n_a(0) - n_{a0}] \quad , \tag{11.1}$$

where n_a and D_a are the electron density and its corresponding diffusion constant inside the intrinsic active region, respectively. The surface recombination velocity is given by [222]

$$S_{\mathrm{v}} = \sigma_{\mathrm{n}} v_{\mathrm{th}} N_{\mathrm{st}} [n_{\mathrm{a}}(0) - n_{\mathrm{p0}}] \quad , \tag{11.2}$$

where σ_{n} is the capture cross-section of the electron by the surface trap state, v_{th} is the thermal velocity of the electrons and N_{st} is the number of surface trap states per unit area.

The surface states in GaAs are known to be formed by surface oxides or an inbalance in stoichiometry. The major direction in trying to eliminate the surface states has been to replace the oxygen atoms at the surface with other group-VI elements, such as sulfur or selenium, to improve both the electrical and optical performances of devices [112, 202, 214, 258]. Instead of the deep donor states for the oxygen, these elements form shallow donor states, and the recombination rate through these states can be significantly reduced. However, the binding energy of oxygen with the GaAs surface is much larger than that of sulfur, and the surface sulfides are eventually replaced by oxides in ambient conditions. Reduction of the surface recombination velocity after treatment in sulfide solution has been reported, but the improved performance only lasted for a short period of time [243]. One solution to this was to cover the sulfur-passivated surface with a silicon nitride film, which was stable for extended periods of time [96, 114]. Such passivation techniques report a reduction in surface recombination velocity of one to three orders of magnitude .

Using the well-known number of $S_{\mathrm{v}} = 10^6$ cm/s [114], the nonradiative recombination lifetime for an electron–hole pair through surface states is estimated to be $\sim$1 ps for sub-micron structures. Since the radiative recombination lifetime is on the order of $1 \sim 10$ ns, this means that there is no way that an electron–hole pair will generate a photon in such a sub-micron structure unless the surface recombination rate is significantly reduced.

We adopted the sulfur-passivation technique followed by silicon nitride encapsulation. First, the etched sub-micron post structures (Fig. 11.2) were dipped into an ammonium sulfide solution [$(NH_4)_2S$] heated to 60 °C and containing some excess sulfur, for 5 min. Then, the sample was rinsed in water and immediately put in a vacuum chamber, where a 500-Å-thick silicon nitride film was deposited using the plasma-enhanced chemical vapor deposition (PECVD) technique. PECVD is a conformal deposition technique, so the silicon nitride film covers the side of the post structure. The PECVD was performed in the same ECR system that was used to perform reactive ion etching, at an extraordinarily low-temperature of 100 °C. Such low temperature deposition is critical for GaAs and was made possible by the high-density plasma generated using the ECR technique.

An experiment was performed to demonstrate the effectiveness of the surface-passivation technique described above. Figure 11.3 shows the current–voltage characteristics of large area (100μm×100μm) p–i–n junctions prepared from identical wafers with different surface preparations, measured at 4 K. The dashed line shows the sample prepared by wet chemical etching in H_3PO_4/H_2O_2 solution. One can see that a significant amount of current

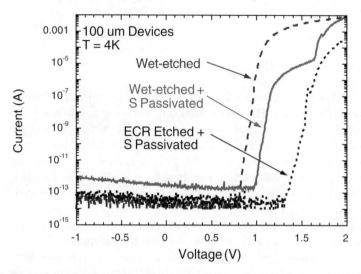

Fig. 11.3. Current–voltage characteristics of p–n junctions with different surface preparations. The *dashed, solid* and *dotted lines* correspond to wet-etched, wet-etched + sulfur-passivated, and ECR-etched + sulfur-passivated samples, respectively

starts to flow at a DC bias voltage of 0.8 V. Since the current should not flow in a p–n junction until the bias voltage reaches the built-in potential ($\sim$1.5 V for GaAs at 4 K), the current below 1.5 V is attributed to the leakage current through the surface. The solid line shows the current–voltage characteristics when this wet-etched device is treated by the sulfur-passivation technique described above. The leakage current decreases by about two orders of magnitude at the voltages below the built-in potential, consistent with other reports. When the surface is prepared by ECR etching and sulfur passivation, the result shown by the dotted line is obtained. The leakage current in this device is significantly reduced until the bias voltage approaches 1.5 V. The surface leakage current at voltages below this value is suppressed by more than 10 orders of magnitude in these samples, compared to the conventional method of wet etching.

The significant reduction of surface leakage current demonstrated in this experiment, compared to reports using simple sulfur passivation treatment, deserves further investigation. One conclusion that can be drawn from this experiment is that it is more important to initially prepare a good surface rather than to try to recover a badly damaged surface in order to suppress the surface recombination. Wet etching of GaAs is achieved by oxidizing the GaAs material with H_2O_2 and removing the oxide with either an acid or a base. That process inherently leaves a high density of surface oxides, which result in a large surface recombination velocity. Dry etching using low energy ECR plasma seems to leave an etched surface with low surface recombination rates. It was also found that annealing the ohmic contacts after the posts have been etched results in high surface recombination rates. This is because either the surface stoichiometry is altered at high temperatures (the GaAs crystal starts losing As atoms at the surface above 400 °C) or oxidation is enhanced at high temperatures. For the samples for which the ohmic contacts are annealed before the p–n junction surface is etched, and the etching is achieved by ECR-RIE, the surface recombination rates were always low. This conjecture is supported by experiments in micron-scale semiconductor microcavity lasers that were demonstrated recently. Assumption of an exceptionally low surface recombination velocity is necessary to explain the threshold values achieved in these devices [196, 199, 211]. The fact that these optically pumped devices did not need any annealing procedure, and that they were also etched using low-energy ECR-RIE or kinetically assisted ion beam etching (KAIBE), is consistent with the conjectures made here to describe our devices.

11.2.7 Planarization and Top-Contact Evaporation

Once the post structure is made and the surface of the p–n junction is properly passivated, the top of the post needs to be electrically contacted to the outside world. In order to achieve this, the structures were first planarized with some insulating material and large contact pads that can be wirebonded were put down. The insulator chosen is hard-baked photoresist, since it is easy to work with and readily available. Photoresist (Shipley 1805 by Microposit) was spun on at 4000 rpm (rotations per minute) to give a $\sim$5000 Å film. Since the thickness of the photoresist film is on the order of the lateral size of the posts, the thickness of the photoresist on top of the posts will be much thinner than 5000 Å when the photoresist is baked due to thermal flow. Since photoresist normally dissolves in acetone and another liftoff process is necessary to put down the contact pads, hard-baking is required to change the chemistry of the photoresist. An independent characterization experiment shows that when photoresist is baked at 150 $\sim$ 180°C for an extended period of time (>12 h in an oven, or >20 min on a hot plate) it becomes resistant to acetone. The electrical characteristics of such hard-baked photoresist is also tested as to whether it is insulating or not. If photoresist is baked at a much higher temperature (above 200°C), the film becomes metallic.

After covering the whole structure with hard-baked photoresist, the top of the posts need to be exposed for further electrical contact. This is achieved by a short plasma etch in O_2 to remove a thin layer of photoresist, and another short plasma etch in CF_4/O_2 to remove the silicon nitride film on top. This process exposes the top metal contact (which was used as the mask to etch the post structures), and another layer of metallization can be added to lead to large wirebonding pads. In this p–i–n junction device, the photons are extracted only through the top, since any photon injected to the substrate side is absorbed by the GaAs substrate. Therefore, the metal contact pad

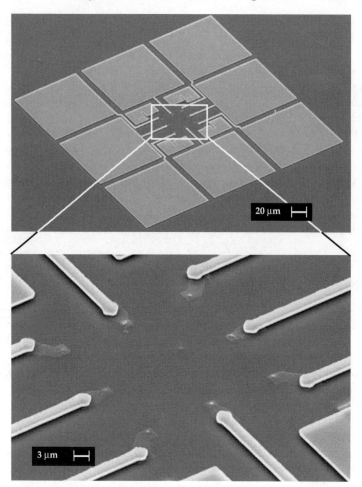

Fig. 11.4. An SEM micrograph of the final sub-micron p–i–n junction device. The large contact pads were 100 μm on one side and used for wirebonding. Eight contact pads are made in one set and are in contact with eight mesoscopic p–i–n junctions. The small pads on top of the posts allow photons to escape

should be thin enough to guarantee sufficient optical transparency. An aligned electron-beam-lithography step was performed to put down thin metal square pads (∼200 Å thick) that align to the top of the posts in one corner. Then, thick (∼3000 Å) gold contact pads were evaporated using optical lithography that was used for wirebonding. A scanning electron microscope (SEM) picture of this is shown in Fig. 11.4.

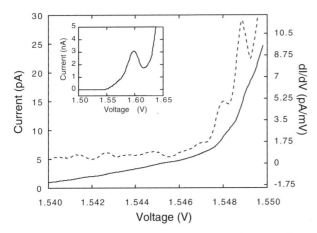

Fig. 11.5. Current–voltage characteristics (*solid line*) and dI/dV–voltage characteristics (*dashed line*) of a higher tunneling rate p–i–n junction at 50 mK. The I–V curve is measured near the lower tail of the resonant tunneling peak. The inset shows the I–V curve for a wider voltage range, in which the resonant tunneling peak can be seen

11.3 Observation of the Coulomb Staircase

In this section, evidence of the Coulomb blockade effect is presented in a mesoscopic p–i–n junction device fabricated using the procedure described above. The p–i–n junction structure studied in this section was slightly different from that described in Table 11.1. The junction used in this device was optimized for higher electron and hole tunneling rates and has lower (15 % Al concentration instead of 18 %) and thinner (160 Å thickness instead of 200 Å) barriers, and the p-side and n-side QWs are made of GaAs instead of AlGaAs. The fabricated devices were installed in the mixing chamber of a dilution refrigerator, at a base temperature of ~50 mK. The junction was biased with DC and AC voltage sources, so that the DC current–voltage characteristics of the device could be measured with the option of an AC modulation signal superimposed on top.

Figure 11.5 shows the DC current–voltage characteristics of a device with a lithographic diameter of 600 nm at a very low bias voltage. The inset shows the measured I–V characteristics over a wider voltage range, in which the resonant tunneling peak can be seen. In this device at a low voltage, the hole tunneling rate curve has a very small slope, which can be derived from the I–V characteristics. As predicted by the model described in Chap. 10, small steps in the I–V curve can be observed, which show up as a periodic oscillation in the calculated dI/dV curve. Each period corresponds to an addition of one electron in the central QW. From the period of approximately 1.1 ± 0.2 mV we derive a junction capacitance of 0.15 ± 0.03 fF.

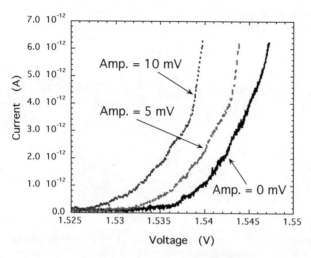

Fig. 11.6. Current–voltage characteristics of the same p–i–n junction as in Fig. 11.5 with an AC square-wave modulation. The frequency was fixed at 10 MHz, and the amplitude was varied to produce the different curves

The Coulomb staircase was also observed at this low bias voltage when a square-wave-modulation signal of frequency $f = 10$ MHz and amplitude ranging from 3 mV to 10 mV was applied. This is shown as the modulation of the DC current as a function of bias voltage in Fig. 11.6. This results from the fact that the modulation frequency is fast compared to the hole tunneling rate. In this situation the current is determined solely by the hole tunneling rate [(10.19) and (10.20)]. As in the case without modulation, the steps in the I–V characteristics reflect the charging of the QW by additional electrons. It was also confirmed experimentally that there was no frequency dependence of the Coulomb staircase steps in this limit.

The current–voltage measurements on this device show that the Coulomb blockade effect exists in a sub-micron p–i–n junction cooled to millikelvin temperatures. This is the first experimental demonstration of the Coulomb blockade effect between two types of carriers (electrons and holes) in a bipolar device.

11.4 Single-Photon Turnstile Device

11.4.1 Preliminary Characterization

Experimental evidence for single-photon turnstile operation was observed in a sample with the junction structure described in Table 11.1. Before searching for experimental evidence for single-photon turnstile device operation, a few measurements were performed to characterize the electrical and optical

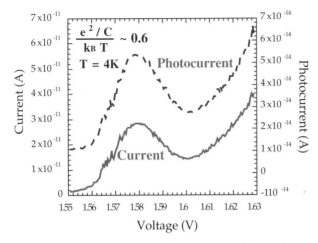

Fig. 11.7. Measured current (*solid line*) and photocurrent (*dashed line*) of a mesoscopic p–i–n junction with a diameter of 600 nm, fabricated by the procedure described in this chapter. The measurement was done at 4 K, and the photocurrent was measured with an avalanche photodiode

performance of the device. These preliminary measurements were done at 4 K.

First, the DC current–voltage characteristics was measured for a device with diameter of 600 nm. A well-defined resonant tunneling current peak with a very small background current was observed, as shown with a solid line in Fig. 11.7. This indicates that surface (leakage) current is suppressed well in spite of the very small size of the post created by the fabrication process described. The maximum value of the resonant tunneling current indicates a maximum electron tunneling rate of 2×10^8 s^{-1}, which is not very far from the estimated value of 7×10^7 s^{-1} calculated in Chap. 10.

The photons that are generated from this were was also measured, using a large area avalanche photodiode (APD; with an active area diameter of ~5 mm). The APD was placed about 1 mm away from the surface of the mesoscopic junction, in a face-to-face configuration. The APD was also cooled down to 4 K, and the avalanche gain was about 60 at the bias point used. The dashed line in Fig. 11.7 shows the measured photocurrent through the APD normalized by the avalanche gain, to provide the primary photocurrent, $I_{\rm p}$. One should note that the shape of $I_{\rm p}$ follows the shape of the device current, $I_{\rm d}$, very closely. The relation between these two currents is given by

$$I_{\rm p} = \eta_{\rm i}\eta_{\rm e}I_{\rm d} \quad , \tag{11.3}$$

where $\eta_{\rm i}$ is the internal quantum efficiency for conversion of an electron–hole pair to a photon and $\eta_{\rm e}$ is the external coupling efficiency for the generated photon to reach the detector. From Fig. 11.7, one can conclude that the overall current-to-photocurrent conversion efficiency ($\eta_{\rm i}\eta_{\rm e}$) is about 2×10^{-3}.

Air : n_1 = 1.0

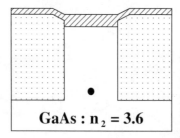

GaAs : n_2 = 3.6

[......] **Photoresist : n_3 = 1.4**

[////////] **Au Metal Contact**

● **Point photon source**

Fig. 11.8. Schematic of a model used to estimate the external coupling efficiency of a photon to escape the device and reach the photodetector

The external coupling efficiency (η_e) for a photon to escape the single photon turnstile device and reach the detector can be estimated from simple geometric considerations. Figure 11.8 shows the model used for this estimation. A model was considered where a point photon source is placed inside a post structure that is surrounded by photoresist and covered by a thin gold film. The thicknesses of the gold film used for the calculation were 500 Å on top of the posts and 300 Å on the photoresist surrounding the structure. The transmittivity of the photons (at λ = 820 nm) through the gold films was calculated [27] to be about 3.8 % on top of the posts and about 16 % on top of the photoresist. Such absorption loss due to the top metal contact is the main reason for reduced total external coupling efficiency. The other reason for low efficiency is the total internal reflection at the GaAs–air interface, due to a large index mismatch. For such a post structure, the total fraction of photons allowed to leave the GaAs crystal is limited to about 9 % due to total internal reflection. Accounting for these two loss mechanisms, the total external coupling efficiency is calculated to be less than $\sim$6 $\times10^{-3}$. From this calculation, a lower bound on the intrinsic electron-to-photon conversion efficiency (η_i) can be estimated to be $\geq$33 %.

11.4.2 Experimental Setup

Figure 11.9 shows the schematic of the experimental setup used in this experiment. The device was installed in a dilution refrigerator with a base temperature of $\sim$50 mK. Both a DC and an AC bias voltage were applied

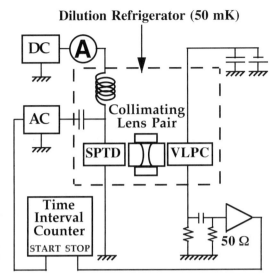

Fig. 11.9. Schematic of the experimental setup for the single-photon turnstile device

using a bias tee, also installed in the mixing chamber, less than 1 inch away from the sample. A DC current flowing through the device was measured as a function of DC bias voltage with a square-wave AC modulation voltage superimposed on top. It is very important to verify that the square-wave AC signal from the function generator is transferred to the sample without significant distortion. The bandwidth of the bias tee used was 100 kHz − 4.2 GHz. Figure 11.10a is the input pulse generated from our pulse generator, with about 45 ns on-pulse duration and 100 ns period. Figure 11.10b shows the pulse measured at room temperature after it passes through the bias network, including the coax cables inside and outside the dilution refrigerator, a 20 dB attenuator installed at the 1 K pot of the dilution refrigerator, and the bias tee placed at the mixing chamber. The input to the oscilloscope was placed about 1 inch away from the output of the bias tee in place of the p–n junction sample. The amplitude of the pulse is decreased by a factor of 10 due to the 20 dB attenuator that was installed to attenuate the room temperature thermal noise and provide good heat-sinking of the AC coax cable. A DC bias of 10 mV was also applied to show that applying DC bias does not affect the pulse shape in this bias tee.

The emitted photons from the mesoscopic p–i–n junction device were focused onto a single-photon counting detector based on a visible-light photon counter (VLPC) using a collimating lens pair. The characteristics of this detector are discussed in detail in Chap. 12. The detector's performance is optimized at a temperature of 6∼7 K. The VLPC was also installed in the mixing chamber of the dilution refrigerator, but the temperature was held at

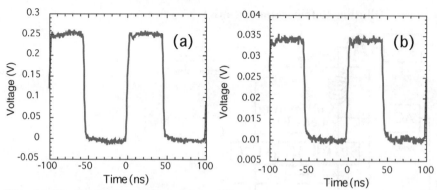

Fig. 11.10 a, b. Real time trace of an AC square wave at (a) the input and (b) the output of the bias network including coax cables inside and outside the dilution refrigerator, a 20 dB attenuator installed at 1 K pot of the dilution refrigerator, and the bias tee placed at the mixing chamber

6.5 K with active temperature control and good thermal isolation provided by eight thin superconducting wires that suspended the detector in vacuum. The detector generates an electrical pulse with a duration of 2 ns and a very well-defined height. This signal was used to trigger the "start" port of a time-interval counter. The "stop" port of the time-interval counter could be triggered either by the rising edge of the AC modulation signal to the p–i–n junction, to measure the correlation between the modulation signal and the output photon flux, or by another photon signal from the photon counter, to measure the intensity–intensity correlation function (or the second-order coherence function) of the output photon flux.

11.4.3 Electrical Characterization

Single-photon turnstile operation requires that the electrons and holes are injected into the active region in a regulated fashion. This means that only one electron and one hole are injected per modulation period and implies that there is a strong correlation between the DC current and the modulation frequency [154]. Figure 11.11 shows the measured current as a function of AC modulation frequency for a device with a diameter of 600 nm. A fixed AC amplitude of 72 mV is superposed on the three different DC voltages, $V = 1.545$, 1.547 and 1.550 V. As the modulation frequency was increased, the DC current increased linearly as a function of the modulation frequency. The measured current was in close agreement with the relation $I = ef$, $I = 2ef$, and $I = 3ef$ (solid lines), when a frequency-independent background current was subtracted. This background current varies from device to device, and ranges from 0.5 pA to 6.5 pA. In Fig. 11.12, the slopes I/f from the current versus frequency curves were evaluated and plotted as a function of the DC bias voltage. It is seen that the slope increases discretely, creating plateaus

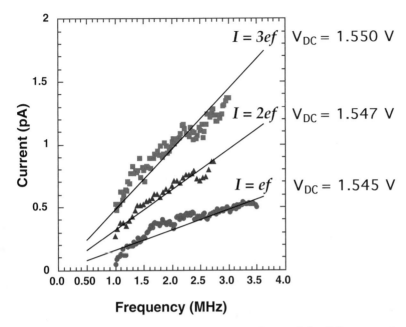

Fig. 11.11. The modulation frequency dependence of the DC current in the 600 nm turnstile device. The background DC (leakage) current, which is independent of the modulation frequency, has been subtracted. The measured current agrees with the relation $I = nef$ (*solid lines*), where e is the electron charge and $n = 1$, 2 and 3

at $I/f = ne$, where $e = 1.6 \times 10^{-19}$ C is the charge of an electron and $n = 1$, 2 and 3.

It is worthwhile mentioning that there is an optimum range ($\sim 72 \pm 5$ mV) for the amplitude ΔV of the AC modulation signal over which turnstile operation is sustained. This arises from the fact that the resonant tunneling in a realistic QW is never ideal, and the peak-to-trough ratio is always finite. If the modulation amplitude is too large, unwanted electron tunneling is allowed at high voltage ($V + \Delta V$) due to other leakage mechanisms, and the fraction of the regulated current decreases. The optimum modulation amplitude (~ 72 mV) is determined by the difference between the resonant tunneling conditions for electrons and holes, while the tolerance (± 5 mV) is constrained by the decreasing ratio of the turnstile current to the (leakage) background current.

The locking of the current at multiples of the modulation frequency ($I = nef$) suggests that the charge transfer through the device is strongly correlated with the external modulation signal [50, 71, 130, 191]. At the first current plateau, $I = ef$, a single (the mth) electron and a single (the first) hole are injected into the central QW per modulation period, resulting in single- photon emission. At the second current plateau, $I = 2ef$, two [the

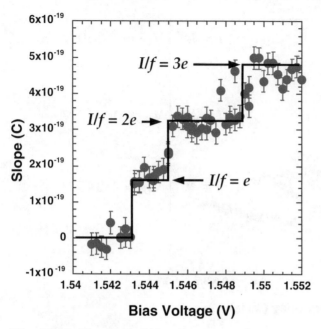

Fig. 11.12. The slopes I/f in the current–frequency curve versus DC bias voltage. The slopes I/f are quantized and forms plateaus at multiples of e, indicating operation at $I = ef$, $2ef$ and $3ef$

mth and $(m+1)$th] electrons and two (the first and second) holes are injected into the central QW per modulation period, resulting in two-photon emission. Similarly at the third current plateau, $I = 3ef$, three electrons and three holes are injected per modulation period, resulting in three-photon emission. This multiple charge operation becomes possible because of relatively broad inhomogeneous linewidths of the n–side and p–side QWs, as discussed in Chap. 10.

11.4.4 Optical Characterization

Having confirmed the regulated injection of electrons and holes into the active region, it is necessary to establish a correlation between the injected carriers and emitted photons, to confirm the turnstile operation for photons. To observe the time correlation between the modulation input and photon emission, we measured the time delay from the rising edge of the modulation input to the photon detection event at the first current plateau ($I = ef$) and the second current plateau ($I = 2ef$). The probability for a single electron–hole pair injected to the central QW of the turnstile device to be detected as a photon in the detector was about 1×10^{-4} due to a poor optical coupling efficiency between the two devices in the present setup. However, the quantum efficiency of detection does not affect the time correlation characteristics.

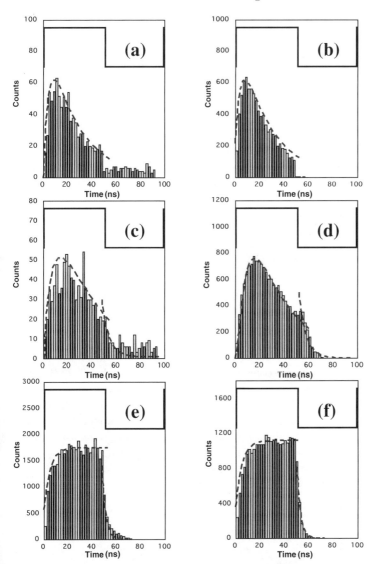

Fig. 11.13 a–f. Photon-emission characteristics of the turnstile device. (**a**) Measured histogram of the time delay between the rising edge of the modulation input and the photon-detection event at the first plateau $(I = ef)$. (**b**) Monte-Carlo numerical simulation result for the first plateau. (**c**) Measured histogram of a time delay at the second plateau $(I = 2ef)$. (**d**) Monte-Carlo numerical simulation result for the second plateau. (**e**) Measured histogram of a time delay for a larger-area device (diameter of 1.4 μm) at higher temperature (4 K), where the Coulomb blockade effect is absent. (**f**) Monte-Carlo numerical simulation result for this modulated classical light-emitting diode case

The histograms of the measured time delay with 10 MHz modulation frequency is shown in Fig. 11.13a (for $I = ef$) and Fig. 11.13c (for $I = 2ef$). The solid line on top indicates the external AC modulation voltage, and the finite photon counts during the off-pulse period are due to the dark counts of the VLPC. For the $I = ef$ plateau, a photon is emitted after the rising edge of the pulse since a single electron is injected at V_0 and a single hole is injected at $V_0 + \Delta V$. The photon-emission probability has a peak near the rising edge of the modulation input. The rapid increase of the photon-emission probability is associated with the hole tunneling time ($\tau_h \simeq 4$ ns), and the slow decay of the photon-emission probability corresponds to the radiative recombination lifetime ($\tau_{ph} \simeq 25$ ns). When only one electron and one hole are allowed to be injected and only one photon is emitted per modulation cycle, the photon-emission probability $P(t)$ for the time t after the turn-on of a higher voltage pulse is given by

$$P_1(t) = \frac{\exp(-t/\tau_{ph}) - \exp(-t/\tau_h)}{\tau_{ph} - \tau_h} \quad , \tag{11.4}$$

where τ_{ph} is the radiative recombination lifetime and τ_h is the tunneling time of the hole. The dashed line is generated by this analytical formula using the parameters τ_h and τ_{ph} found from the experiment. Photon-emission probability in Fig. 11.13a decays to a non-zero value during the on-pulse due to photons generated by background current. The ratio of the counts contained in the peak to those contained in the non-zero background is $\sim$3:1, consistent with the ratio of the turnstile current to the background current in this device. Figure 11.13b shows the Monte-Carlo numerical simulation performed for this situation; it reproduces the experimental result well.

For the $I = 2ef$ plateau shown in Fig. 11.13c, the photon-emission probability distribution is broader, since two electrons and two holes are injected, and two photons are generated per modulation period. The photon-emission probability for the case where two electrons and two holes are allowed to tunnel can also be calculated analytically, as

$$P_2(t) = \frac{2\tau_{ph} - \tau_{h2}}{2(\tau_{ph} - \tau_{h1})(\tau_{ph} - \tau_{h2})} \exp(-t/\tau_{ph})$$
$$+ \frac{2\tau_{h1} - \tau_{h2}}{2(\tau_{h1} - \tau_{h2})(\tau_{h1} - \tau_{ph})} \exp(-t/\tau_{h1})$$
$$+ \frac{\tau_{h2}}{2(\tau_{h2} - \tau_{h1})(\tau_{h2} - \tau_{ph})} \exp(-t/\tau_{h2}) \quad . \tag{11.5}$$

Here, τ_{h1} is the tunneling time for the first hole and τ_{h2} is the (longer) tunneling time for the second hole. This theoretical curve is shown as a dashed line in Fig. 11.13c. The sharp cutoff of photon emission after the falling edge of the modulation input is caused by the decay of the hole population due to simultaneous radiative recombination and reverse hole tunneling. The associated lifetime for this decay is $(\tau_h^{-1} + \tau_{ph}^{-1})^{-1}$. The experimental results as well as the analytical traces are well reproduced by the Monte-Carlo numerical simulation, as shown in Fig. 11.13d.

In order to compare this with photons from a classically modulated light-emitting diode, a control experiment is necessary where the Coulomb blockade effect is absent. Figure 11.13e shows the measured histogram of a time delay for a larger-area device (diameter of 1.4 μm) at higher temperature (4 K). In this case, an arbitrary number of holes are allowed to tunnel into the central QW during an on-pulse, and so the resulting photon-emission probability should increase monotonically to a steady-state value. This result is well-reproduced by the simulation, as shown in Fig. 11.13f. An analytical solution for the photon- emission probability can be found, and is given by

$$P_3(t) = \frac{C}{\tau_h} [1 - \exp(-t/\tau_{ph})] \quad , \tag{11.6}$$

where C is a normalization constant. This analytical solution was used to generate the dashed lines in Fig. 11.13e and f. The photon-emission probability increases monotonically through the duration of the on-pulse with the time constant τ_{ph}, as one would expect from a turn-on of a classical photon source. The radiative recombination time is shorter in this case ($\simeq$5 ns) due to a higher carrier density under the operating conditions.

The photon emission probabilities shown in Fig. 11.13a–d and (11.4) and (11.5) are in sharp contrast to an ordinary pulse-driven classical photon source where the Coulomb blockade is absent [shown in Fig. 11.13e and f and (11.6)]. The fact that the photon-emission probability decreases during the on-pulse duration is a unique signature that the number of holes injected during an on-pulse is restricted to either one or two due to the Coulomb blockade effect. This is a strong evidence for the Coulomb blockade effect and turnstile operation of our mesoscopic p–n junction.

11.5 Summary

In this chapter, the experimental effort towards generation of heralded single-photon states using the Coulomb blockade effect in a mesoscopic p–i–n junction was discussed. The process used to fabricate these small junctions was described, and the results of measurements performed at low temperatures were presented. The first experimental observation of a Coulomb staircase was reported in a system involving two different types of particles in a bipolar device.

Experimental evidence was also presented for the operation of a single-photon turnstile device. The electrical characterization demonstrates a frequency dependent current, which is a signature of time-correlated charge transport due to the Coulomb blockade effect. The optical characterization, where we measured the photon-emission probability distribution with respect to the external modulation signal, implies that the number of carriers that are injected, and thus the number of photons that are emitted, are restricted due to the Coulomb blockade effect.

Such experimental evidence provides proof of the fact that the Coulomb blockade effect can be utilized to manipulate photon statistics and produce sub-Poissonian and antibunched photon states even down to a single-photon level. However, there are imperfections in the current devices that prohibit a direct demonstration of these nonclassical features. Improvements on the device performance is possible, with better confinement of both carriers and photons [90, 91].

12. Single-Photon Detection with Visible-Light Photon Counter

12.1 Introduction

The main focus of the discussion in this chapter is the detection of single photons. The ultimate goal for sensitive photodetection would be the detection of single photons with high quantum efficiency (QE) and low error probability. Experimental techniques for single-photon detection have made tremendous progress in recent years. Successful development of such a sensitive photodetector will be useful for various optical precision measurements like spectroscopy and interferometry. Besides these conventional applications, the development of a high-performance single-photon counter has been crucial for recent progress in quantum information technology utilizing single photons [215], such as quantum cryptography [35, 176, 177], quantum computation [174, 236] and quantum teleportation [28, 67], and fundamental tests of quantum mechanics such as a loophole-free test of Bell inequality experiment [43, 44, 6, 59, 135, 204].

An ordinary photodiode cannot generate a large enough electrical signal from a single detected photon to overcome the huge thermal noise generated in the following electronic circuits; thus, a photodetector with an internal gain mechanism is required. Photomultiplier tubes (PMTs) and Si avalanche photodiodes (APDs) [47] have been most widely used, while alternate technologies like superconducting tunnel junctions (STJs) [185], solid state photomultipliers (SSPMs) [188], and visible-light photon counters (VLPCs) [8, 9] have recently demonstrated unique capabilities that PMTs and APDs cannot offer.

This chapter starts with an overview of the merits and demerits of the various single-photon-counting detectors available at present. The performance characteristics of a single-photon-counting detector based on a VLPC is the main topic of this chapter. The operation principle of this device and the detection system that was developed, including the cryostat and the electronics that optimize the performance, are presented.

The noise properties of the internal gain mechanism determine the distribution of pulse heights generated by single-photon detection events [237]. For detectors with low multiplication noise, the pulse height originating from a single-photon detection event is well-defined. For such detectors, the pulse-height distribution features a narrow dispersion and is separated from the background noise of the electronics; it is very easy to distinguish the photon pulse from electronic noise using a simple voltage discriminator [124]. This is not the case for a detector with large multiplication noise. It is desirable to have a single-photon-counting detector with minimal multiplication noise, so that the discrimination of signal pulses can easily be achieved. Furthermore, when the heights of pulses generated by a single-photon detection event are well-defined, there is a strong correlation between the detected photon number and the height of the resulting pulse. This opens up a possibility to distinguish a single photon detection event from a two-photon detection event [126].

PMTs have low noise in the multiplication process [26], but the single-photon QE is limited to <25 % and so the maximum two-photon detection efficiency should be only 6%. High QE can be achieved by APDs (∼76 %) [133], but the large multiplication noise in these devices completely washes out the correlation between the number of incident photons and the generated pulse height [169, 255]. In addition to that, for the APDs that operate in the Geiger mode for single-photon counting, the entire diode breaks down upon detection of a single-photon incidence. Therefore, it is impossible to distinguish a two-photon detection event from a single-photon detection event.

In the second half of this chapter, we present the results of an experiment in which the noise in the multiplication process, characterized by the excess noise factor (ENF), of a VLPC is measured. A VLPC features noise-free avalanche multiplication, so that the pulse height resulting from a single-photon detection event is extremely well defined. This allows one to correlate the pulse height to the number of photons detected. An experimental result is presented where a two-photon detection event is distinguished from a single-photon detection event with a high QE and low bit error rate.

12.2 Comparison of Single-Photon Detectors

Table 12.1 compares the performances of the detectors available for single-photon detection.

12.2.1 Photomultiplier Tubes (PMTs)

A PMT converts a photon into an electron from a photocathode at the input. This initial electron, which is injected into a vacuum, is accelerated towards a dynode, where more electrons are generated. These electrons multiply at successive dynodes, until the total number of electrons is large enough ($10^6 \sim 10^8$) to be detected as an electrical signal. The QE of the initial photon to generate an electron is limited to about 10 % in the visible and near infrared (IR), and can be as high as 25 % in the ultraviolet. The ENF, which characterizes the fluctuation in the gain, is small (∼1.2) [26], since the number of multiplication events is defined by the number of dynodes in the system. The only

Table 12.1. Performance comparison of various single-photon-counting detectors. M: multiplication gain; QE: quantum efficiency

	Photo-multiplier tube (PMT)	Avalanche photodiode (APD)	Superconducting tunnel junction (STJ)	Visible-light photon counter (VLPC)
Gain process	Electron multiplication at dynodes	Avalanche breakdown of p–n junction	Breaking of Cooper pairs into electrons	Impact-ionization of impurity band
Gain	$10^6 \sim 10^8$	$10^2 \sim 10^6$	$\sim 10^3$	$\sim 3 \times 10^4$
QE	≤ 25 %	≤ 76 %	≤ 45 %	≤ 88 %
Excess noise factor	~ 1.2	$\sim M$	~ 1.2	~ 1.0
Time response	0.3 ns	50 ns	2 ns	2 ns
Dark count	<100 cps	250 cps	Negligible	20000 cps
Detection wavelength	UV-visible	<1 μm	<700 nm	<1 μm
Multi-photon detection	possible	not possible	possible	possible
Energy resolution	No	No	Yes	No

source of fluctuation is the number of electrons emitted from each dynode per input electron. Since the acceleration of electrons in a vacuum can be fast, electrical pulses of widths as narrow as ~ 300 ps can be generated.

12.2.2 Avalanche Photodiodes (APDs)

An APD utilizes avalanche breakdown of a reverse-biased p–n junction as the multiplication process. Most APDs used for single-photon detection operate in a Geiger mode [47], although there are reports of cooled APDs used in a normal bias condition below the breakdown voltage [255]. In the Geiger mode, an APD is biased with a reverse bias voltage above the avalanche breakdown voltage, but the breakdown current is suppressed with a series resistance. When a photon is absorbed at the depletion region of the junction, a single electron–hole pair is generated. The electron and the hole are accelerated in opposite directions in the conduction and the valence band, respectively. When they acquire enough kinetic energy, they impact-ionize, creating another electron–hole pair. This multiplication process triggers an avalanche of the junction that allows a huge current to flow, which forms the electrical pulse that is detected. In order to reset the detector to the "ready" state, the avalanche current must be quenched. In passive quenching, this is done by a series resistor, which causes the bias voltage across the junction to decrease when there is a large current flowing through the resistor. Faster quenching can be achieved using an active quenching technique. In active quenching, the current through the junction is monitored, and when it exceeds a certain value, an active bias circuit reduces the output voltage of the source that biases the junction to below the breakdown voltage. The avalanche is quenched quickly, and the detector is ready for the next photon detection. The QE of an APD can be very high, and a value of ∼76 % was reported a few years ago [134, 133]. The avalanche multiplication process in an APD is known to be very noisy, with an ENF roughly proportional to the gain itself [169]. One drawback of APD for single photon detection is that it can generate dark counts due to after pulsing. Afterpulsing is where a carrier trapped in a trap state in the depletion layer of a junction during avalanche breakdown is released after the avalanche current is quenched to generate another breakdown. There also is a dead time between photon-detection events, which, even with the most sofisticated active quenching technique, is on the order of ∼ 50 ns.

12.2.3 Superconducting Tunnel Junctions (STJs)

STJs have been used to detect photons in the X-ray part of the spectrum for some time. Recent progress in the fabrication and measurement techniques has enabled single-photon detection in the visible. When an STJ is voltage biased below the superconducting gap, no current flows because of the absence of the density of states at the Fermi level in a superconductor, provided that the temperature is low enough. When a photon is absorbed at one terminal of the tunnel junction, it excites a quasiparticle far above the Fermi level, by an amount corresponding to the energy, $\hbar\omega$, of the photon. If this quasiparticle is confined well and allowed to equilibrate in the superconductor, it loses its kinetic energy by breaking other Cooper pairs. When this process is over, the superconductor will have roughly $N = 2\hbar\omega/\Delta$ quasiparticles, where Δ is the superconducting energy gap that corresponds to the binding energy of a Cooper pair. These quasiparticles will tunnel across the junction at finite bias, and the amount of tunnel current can be measured to determine a photon-absorption event. The advantage of this detector is that the number of quasiparticles generated due to a photon-absorption event is proportional to the energy of the photon, which provides a finite energy (wavelength) resolution of ∼40 nm. The highest QE of this type of detector reported so far is ∼45 %, although an optimized device is expected to give a value that is much higher than that. The ENF of this detector is quite small and is similar to that of a PMT (∼1.2). The operation temperature of the device has to be far below the superconducting transition temperature, since the thermally excited quasiparticles provide a fundamental source of noise.

12.2.4 Solid-State Photomultipliers (SSPMs) and Visible-Light Photon Counters (VLPCs)

A SSPM utilizes impact-ionization of carriers between a shallow impurity band and the conduction band as a multiplication mechanism [188]. It has an intrinsic capability to detect single photons with wavelengths as long as 30 μm. A VLPC is a variation of an SSPM, where the photon-absorption layer is separated from the gain region to optimize the sensitivity to visible photons [8, 216, 237]. The performance characteristics of a VLPC are the topic of Sects. 12.3 to 12.5. To summarize, a QE of 88 % has been obtained for a VLPC. The ENF has been measured to be ~1.026, which demonstrates the noise-free avalanche multiplication present in these devices. A pulse with 2 ns can be achieved with a wide bandwidth readout circuit, and a good signal-to-noise ratio (SNR) can be achieved since the internal gain is large (~ 30000). One drawback of this detector is a high dark-count rate, on the order of ~20000 counts per second (or cps), which is initiated by the direct tunneling of electrons from the impurity level to the conduction band.

As far as the desired characteristics of a single-photon counting detector is concerned, a VLPC has demonstrated the best performance characteristics in terms of QE and multiplication noise. The only significant drawback is its high dark-count rate, which may be an important factor in some applications.

12.3 Operation Principle of a VLPC

Figure 12.1 shows the schematic structure of a VLPC made out of silicon. The structure consists of an intrinsic photon-absorption layer a moderately doped gain layer, and a heavily doped contact layer. The electrical contacts are provided at the substrate through the contact layer, and on top by a transparent contact. The top of the detector has an antireflection coating, optimized for photons at 550 nm.

A single photon can be absorbed either in the intrinsic absorption layer, or in the gain layer that is doped with As. When a photon is absorbed within

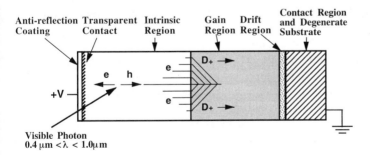

Fig. 12.1. Schematic structure of a visible-light photon counter (VLPC)

the absorption layer, an electron–hole pair is generated. Due to the applied bias, the electron drifts towards the top contact, while the hole drifts towards the gain layer. The gain layer is moderately doped with As, which forms an impurity energy level 54 meV below the bottom of the conduction band in Si. At an operating temperature of 6~7 K, the electrons do not have enough thermal energy to become excited into the conduction band, and are frozen out on the impurity atoms. The incoming hole has a finite cross-section to impact-ionize an electron from the donor level into the conduction band. The initial electron is subsequently accelerated toward the top contact, and impact-ionizes other electrons from the impurity As atoms. A small avalanche breakdown is formed between the impurity band and the conduction band. For the photons that are absorbed in the gain layer, the photoexcited electron plays the role of the initial electron and an identical avalanche is generated. The electrons generated by the avalanche multiplication process drift toward the top contact on a time scale on the order of a nanosecond and provide an electric pulse that is used as the photon signal. When the electrons are excited into the conduction band from the impurity atoms, they leave a trail of ionized impurity atoms behind. The doping level of the gain layer is adjusted so that the electronic wavefunctions of the impurity atoms overlap slightly, forming a narrow impurity band. The ionized donor atoms correspond to holes in this impurity band (usually denoted as $D+$ charges). The doping level allows a slow hopping conduction mechanism in the impurity band, and the impurity-band holes relax towards the contact layer through this mechanism. This impurity-band hole relaxation is a slow process that occurs on a time scale on the order of a microsecond. Until this recovery is complete, the avalanche multiplication area is not available for a second photon detection.

An SSPM is a device that is made entirely of doped layers. IR photons with long wavelengths can photoexcite a carrier, so the sensitivity of SSPM to IR photons is very high. By separating the photon-absorption region from the gain region, the sensitivity of the VLPC to IR photons (of wavelength $1\,\mu m \leq \lambda \leq 30\,\mu m$) is decreased to about 2 %, so that the sensitivity to visible photons is maximized.

12.4 Single-Photon Detection System Based on a VLPC

Impact-ionization of shallow impurity atoms provides the mechanism of avalanche multiplication in a VLPC. Electrons in the impurity atoms can be directly photoexcited into the conduction band, and these electrons can trigger an electrical pulse that is identical to those caused by the valence-band hole generated by an optical photon. Although this direct photoexcitation of the impurity atoms is suppressed by separation of the absorption region and the gain region, a VLPC is still sensitive to long wavelength photons $(1\,\mu m \leq \lambda \leq 30\,\mu m)$. This means that a large number of thermal photons present in this wavelength range can cause huge dark counts in a VLPC. The

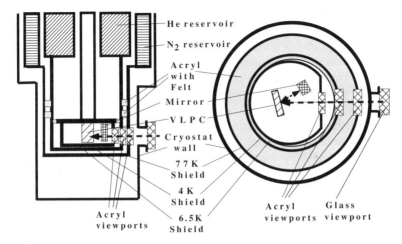

Fig. 12.2. Schematic of the cryostat system with a threefold thermal radiation shield designed for the VLPC

most important aspect of developing a single-photon detection system using a VLPC is to suppress such dark counts caused by the room-temperature thermal photons.

Since the number of thermal photons contained in the spectrum between 1 μm and 30 μm is very large, it is impossible to use a conventional optical window (made out of quartz or fused silica) that has a high transmittivity to photons with wavelengths up to 2∼3 μm. In most applications, the VLPC is installed inside a photon-tight cryostat that is perfectly shielded from both the signal photons and the room-temperature thermal photons. The signal photons are coupled into the detector by means of a plastic fiber [133]. The acrylic material used for the fiber strongly attenuates photons with wavelengths over ∼1 μm; thus, the leakage of thermal photons can be completely eliminated in this scheme. The drawback is that there is a finite reflection of photons between the fiber and the VLPC surface, which causes the coupling efficiency to drop. Kwiat et al. [133] used a spherical refocusing mirror at one end of the fiber to recollect the reflected photons, but even with this addition, the coupling efficiency was below 75 %. They observed an overall QE of ∼70 %, but estimated that the internal QE could be as high as ∼93 % if such reflection loss was accounted for.

Figure 12.2 shows a schematic of the cryostat developed for the VLPC. A bath-type cryostat was used because of the temperature stability and long hold time of He. A small amount of liquid He was drawn to the sample cold plate through a needle valve, and the temperature of the cold plate was actively controlled by a heater. Acrylic optical windows with an antireflection coating were used to couple signal photons into the detector from outside.

Threefold radiation shields at 77 K, 4 K, and the operation temperature of 6.5 K were used to reduce the number of thermal photons reaching the detector. The windows are properly heat-sunk to the shields at appropriate temperatures and act as a perfect cold filter for thermal photons. Room-temperature thermal radiation is estimated to be attenuated by more than 14 orders of magnitude by the shielding system, while the transparency of the total window system was maintained at around 97 % by an antireflection coating at the measurement wavelength of 694 nm. The antireflection coating of the VLPC itself was not optimized for the measurement wavelength, and there was a residual reflection of ~15.6 % at the surface of the VLPC. This was recollected using a spherical refocusing mirror with 99 % reflectance installed inside the cryostat. By using this technique, the total reflection loss at the surface of the VLPC was reduced to below 2.5 %.

For the electrical readout, two different schemes were used. In the first scheme, a custom-made low-temperature amplifier with a bandwidth of about 40 MHz was used. The amplifier utilizes a GaAs MESFET (metal–semiconductor field-effect transistor) as the first amplification stage located near the sample at low temperature, in a cascode configuration with a silicon JFET (junction field-effect transistor) at room temperature [145, 146]. The amplifier response was slow, and an electrical pulse of ~25 ns duration was produced for a photon-detection event. Since the measurement bandwidth was low, it was easier to shield the cryostat electrically from the electromagnetic interference noise. In the second and improved scheme, a low-input-noise silicon amplifier (MITEQ AUX 1347) operating at room temperature was used. The signal from the detector was extracted by a coax cable impedance matched at the input of the amplifier, which acted as a simple 50 Ω resistive load. In this configuration, the bandwidth of the readout circuit was mainly determined by the capacitance of the VLPC (~14 pF) and the 50 Ω input impedance of the amplifier to be about 500 MHz. The resulting electrical pulses were 2 ns in duration, and a good SNR against the electrical noise background was maintained.

The first scheme was used in the QE measurement described in Sect. 12.5 and in the analog noise measurement of the multiplication process described in Sect. 12.7. The second scheme was used in the two-photon detection experiment discussed in the Sect. 12.8.

12.5 Quantum Efficiency of a VLPC

The QE of the single-photon detection system described in the previous section was measured. Continuous-wave (CW) light at 694 nm dfrom a laser diode was attenuated by neutral-density (ND) filters and directly focused onto the VLPC from outside the cryostat through the acrylic windows. The number of incident photons N_{in} (per second) was estimated by the relation $N_{in} = P\lambda/ch$, where λ is the wavelength, c is the speed of light, h is Planck's

constant, and P is the power of the light incident on the VLPC. P is estimated by $P = \alpha P_0$, where P_0 is the power of the diode laser output measured
by a power meter (Coherent Fieldmaster with LM-2 head) with 3 % uncertainty, and α is the total attenuation provided by the ND filters with 1 %
uncertainty. The total attenuation was calculated by simply multiplying the
measured attenuation of each ND filter.

Once N_{in} is known, the QE of the total system can easily be measured
by counting the pulses at the output of the amplifier. Since a slow amplifier
was used in this measurement, dead-time effects due to a slow response of
the electronic circuit need to be taken into account. This is because a pulse-
counting circuit counts pulses above a certain discriminator level, and if a
second photon arrives before the output voltage from the first pulse has relaxed to below this value, it will not be counted. This dead time, τ, caused by
the slow response of the circuit can be estimated from a saturation measurement to be, $\tau \simeq 100$ ns. If a count of N pulses were recorded by the pulse
counter, the actual number of pulses, N', generated by the detector circuit
is estimated to be

$$N' = \frac{N}{1 - N\tau} \quad . \tag{12.1}$$

Therefore, the overall QE η_{tot} of the whole detector system is given by

$$\eta_{tot} = \left(\frac{N_S}{1 - N_S\tau} - \frac{N_B}{1 - N_B\tau} \right) / N_{in} \quad , \tag{12.2}$$

where N_S is the observed counts with input signal photons and N_B is the
observed dark counts without input photons. The observed counts carry 1 %
standard deviation.

Figure 12.3 shows the QE of the detection system as a function of operation temperature and bias voltage across the VLPC. The highest QE of
$88.2 \pm 5\%$ was observed with a bias voltage of 7.3 V at 6.91 K. This observed
value is the highest ever reported for a single-photon-counting QE. In order
to estimate the intrinsic QE of the VLPC itself, one needs to account for
the optical losses that are still present in the setup. First is the finite reflectivity of the optical windows, which is estimated to be 3.1%. Then, there is
residual reflection loss at the surface of the VLPC even with the refocusing
mirror, which is about 2.4 %. There can also be loss due to the transverse
mode profile of the laser beam. When the power of the laser was measured
through a pinhole of 1.0 mm diameter at the focal point of the focusing lens,
1.5 % of the photons were found to be focused outside the sensitive area of
the VLPC. With these corrections, the internal QE of the VLPC is estimated
to be $94.7 \pm 5\%$ at a wavelength of 694 nm.

From Fig. 12.3, one can also see that the QE decreases as the temperature
and the bias voltage are decreased. At low bias, only a small fraction of
electrons generated by photon absorption in the gain layer gain enough kinetic
energy from the electric field to initiate an avalanche breakdown before they

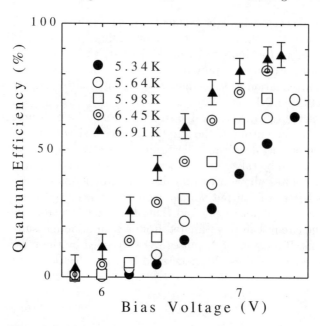

Fig. 12.3. Measured quantum efficiency (QE) of the single-photon detection system as a function of the bias voltage across the VLPC at various temperatures. The error bars are omitted for simplicity, except for the 6.91 K data

move into the undoped layer. At higher bias voltages almost all gain-layer-generated electrons can attain sufficient energy to initiate avalanches. This is the main degradation mechanism for the QE at low bias voltages. The QE versus bias voltage curve shifts to a higher voltage as the temperature is decreased. This is because the deepest part of the gain layer acts like a large series resistor, which becomes more resistive as temperature is decreased. This series resistance reduces the effective bias voltage applied to the active part of the gain layer where the avalanche is produced. When the bias voltage was increased at temperatures above 7 K, the QE reached its highest value (88 %) at a lower bias voltage (compared to 7.3 V) and saturated. When the bias voltage across the device was increased above 7.3 V, the device breaks down, and a huge current starts to flow. The breakdown voltage did not depend strongly on temperature. It is caused by the direct tunneling of electrons from the bound As atoms to the conduction band, which initiates avalanche breakdown.

Figure 12.4 shows the dark counts of the VLPC plotted as a function of the QE, measured at various temperatures and bias voltages. The dark-count rate was 20000 cps at 6.91 K with a bias voltage of 7.3 V, when the maximum QE was observed. Less than 500 cps of the dark counts were caused by the leaking room-temperature thermal radiation, as was confirmed by covering the viewports on the radiation shields. Most of the observed dark counts were

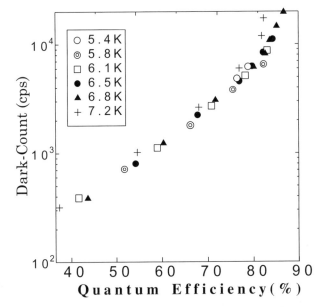

Fig. 12.4. Dark counts of the VLPC plotted as a function of the QE, measured at various temperatures and bias voltages

caused by the internal mechanism of the detector. The data show that the measurements performed at various bias voltages and temperatures collapse onto one universal curve. The dark counts of the VLPC increased roughly as the exponential of the QE when the temperature or the bias voltage was modified.

The saturation effect of the detector response was also measured as the input photon number N_{in} was varied, as shown in Fig. 12.5. The spot size of the laser beam at the detector surface was optimized to give a maximum count rate when $N_{in} = 2 \times 10^5$ (crosses). When the spot size was too small, the count rate decreased due to detector saturation caused by slow impurity-band hole ($D+$) relaxation (solid circles). When the spot size was too large, the count rate decreases because the photons contained in the tail of the transverse spatial mode missed the sensitive area of the VLPC (open diamonds). The optimum spot size can be estimated by scanning the detector laterally and measuring the point at which the count rate starts to decrease significantly. At this input photon flux, the optimum spot size was about 100 μm (crosses). When N_{in} was varied from 10^4 to 2×10^7, (12.1) fitted well to the measured counts with $\tau = 100$ ns. This saturation is caused purely by the readout circuit and pulse counting circuitry. When N_{in} was smaller than 2×10^5, the QE degradation was less than 1 % due to this effect.

One can also estimate the QE degradation due to the intrinsic recovery time of the VLPC. Figure 12.5 shows that the QE is degraded to $\frac{1}{2}$ (3 dB)

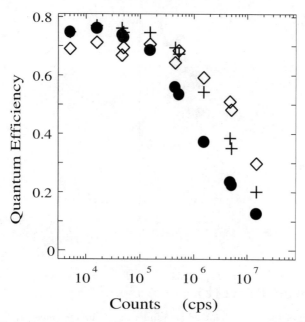

Fig. 12.5. QE of the VLPC measured as a function of the incident photon number. *Solid circles*, *crosses* and *open diamonds* correspond to increasing spot size of the input beam. The spot size was about $100\,\mu$m in diameter for the crosses

when the input photon flux is $\sim$5 $\times 10^6$ (for crosses). When a detected photon creates an avalanche breakdown and generates a pulse, that local area cannot support another multiplication process until the impurity-band holes completely recover by hopping conduction. This process takes $\sim$2 µs. The estimated area of an avalanche breakdown in a VLPC has a radius on the order of the thickness of the gain layer, $\sim$15 µm. If we interpret an area with a diameter of 30 µm as an independent detector with a dead time of 2 µs, there are about 10 such independent detectors in the 100 µm spot of the laser beam. The 3 dB saturation point is expected if each independent detector receives $1/(2\,\mu\mathrm{s}) = 5 \times 10^5$ cps of input photons. This corresponds to an input photon flux of 5×10^6, which is consistent with the value obtained from Fig. 12.5. It should be noted that at a given N_{in} the spot size can be optimized somewhat to maximize the QE. For example, when $N_{\mathrm{in}} = 10^6$, it was possible to minimize the QE degradation to 7% by an optimum choice of the spot size.

12.6 Theory of Noise in Avalanche Multiplication

12.6.1 Excess Noise Factor (ENF)

This section starts with a brief description of the theoretical analyses on noise in the avalanche multiplication process. There have been many theoretical studies on this issue [85, 169, 234]. The noise property of the avalanche multiplication process is characterized by the ENF, F, defined by

$$F \equiv \frac{\langle M^2 \rangle}{\langle M \rangle^2} \tag{12.3}$$

where M is a statistical variable that describes the multiplication gain, and the angled brackets denote statistical (ensemble) averaging. $F = 1$ is the theoretical limit for low F, achievable in an ideal noise-free avalanche multiplication where each and every primary carrier generates exactly M carriers as a result of the multiplication.

The first theoretical calculation of the ENF in an avalanche photodiode was presented in a seminal paper by McIntyre in 1966 [169]. He analyzed a situation where electrons and holes subject to a large electric field in the depletion region of a reverse-based p–n junction generate another electron–hole pair by impact-ionization. The electron and hole ionization coefficients (ionization probabilities per unit length) are called α and β, respectively. An important assumption made in this analysis is that the coefficients α and β are functions of the electric field only, and do not depend on the history of the carrier. The implication of this assumption is that the electron or hole that is generated by impact-ionization gains enough kinetic energy in a very short time and is immediately available for impact-ionization of another electron–hole pair. Such a "Markovian" approximation is valid in a conventional avalanche photodiode, where the electric field in the depletion region is large and the carrier acceleration is large.

Figure 12.6 shows the calculated result for the ENF as a function of the multiplication gain [38, 169]. $k = \beta/\alpha$ is the ratio between the ionization coefficients of holes and electrons. This calculation result shows that the lowest ENF achievable in an APD is $F = 2$ when only one type of carrier (either an electron or a hole) can impact-ionize another electron–hole pair. The residual noise stems from the fact that the number of impact-ionization events that occur in multiplying a single input carrier is still random, since the ionization process is a random point process with a constant probability (α or β) per unit length. When the ionization coefficients are equal ($\alpha = \beta$), the ENF is equal to the multiplication gain ($F = M$). For a real Si APD used for single-photon detection, the electron and hole ionization coefficients are roughly the same, and the ENF is roughly proportional to the gain itself.

In order to achieve a photon detector with low multiplication noise, it is important to increase the ratio between the electron and hole ionization coefficients. However, the lowest ENF achievable by this approach has a limit

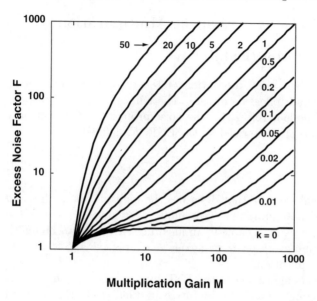

Fig. 12.6. Excess noise factor (ENF) of the avalanche multiplication process in an avalanche photodiode as a function of the multiplication gain. k is the ratio between the ionization coefficients of holes and electrons. The curves are generated from the theoretical expression derived in [169]

of $F = 2$, unless one finds a mechanism where the Markovian assumption is not valid. There have been several experimental attempts to regulate the randomness in the impact-ionization process. One possibility is to use the space-charge effect [5] to modify the electric field as the avalanche multiplication proceeds. The other is to design a structure where the spatial location of the impact-ionization event is localized [37, 131, 247], as in the case of a PMT.

12.6.2 Noise Power Spectral Density of the Multiplied Photocurrent

The noise power spectral density of the photocurrent, generated by the photodetector with the multiplication gain in response to an input photon flux, can be described as [85, 234]

$$S_{\Delta I} = 2e\langle I_{\mathrm{ph}}\rangle\langle M\rangle^2[F_{\mathrm{ph}} + (F - 1)] \quad . \tag{12.4}$$

Here, $\langle I_{\mathrm{ph}}\rangle$ is the average input photocurrent before multiplication (primary photocurrent), $\langle M\rangle$ is the average gain and $F_{\mathrm{ph}} = \langle \Delta n_{\mathrm{ph}}^2\rangle/\langle n_{\mathrm{ph}}\rangle$ is the Fano factor of the incident photons, where $\langle n_{\mathrm{ph}}\rangle$ and $\langle \Delta n_{\mathrm{ph}}^2\rangle$ are the mean and variance of the input photon number. When the incident light is shot-noise limited, $F_{\mathrm{ph}} = 1$ and the above expression reduces to

$$S_{\Delta I} = 2e\langle I_{\mathrm{ph}}\rangle\langle M\rangle^2 F$$
$$= 2e\langle I_{\mathrm{ph}}\rangle(\langle M\rangle^2 + \sigma_{\mathrm{M}}^2) \quad , \tag{12.5}$$

where $\sigma_{\mathrm{M}}^2 = \langle M^2\rangle - \langle M\rangle^2$ is the variance in the multiplication gain. Interpretation of (12.5) is simple: the total noise power spectral density of the multiplied current should be the sum of noise in the primary photocurrent (before amplification) multiplied by the average gain squared (since this is power) and the noise added in the multiplication process, described by the variance. When the multiplication process is noise free ($F = 1$), then it is simply given by the primary photocurrent multiplied by the square of the average gain.

12.6.3 Effect of ENF in the Pulse-Height Distribution

If the avalanche multiplication process is completely noise free ($F = 1$), one can imagine that the number of carriers released at the output photocurrent due to a single-photon detection event is exactly M, where M is the (average) multiplication gain. The heights of the pulses corresponding to single-photon detection events, determined by the number of carriers produced in the multiplication process, are also extremely well defined and feature a delta-function-like pulse-height distribution. As the noise in the multiplication process is increased, the heights of the pulses will start to fluctuate, and the pulse-height distribution gains finite width. The theoretical expression for the pulse height distribution is given by the gamma distribution [237]:

$$P(M) = \frac{1}{M}\left(\frac{1}{F-1}\frac{M}{\langle M\rangle}\right)^{\frac{1}{F-1}} \frac{\exp\left(-\frac{1}{F-1}\frac{M}{\langle M\rangle}\right)}{\Gamma\left(\frac{1}{F-1}\right)} \quad . \tag{12.6}$$

Figure 12.7 shows the pulse height distribution when the average gain $\langle M\rangle = 20000$, for various values of F. One can see that when F is very close to one, the distribution features a sharp peak with a narrow dispersion. The center of the peak is well separated from 0, and observation of a pulse with this height indicates a single-photon detection. As the ENF increases, the distribution broadens and the center of the peak decreases towards 0. When $F = 2$, as in the ideal avalanche photodiode case, the distribution becomes an exponential distribution, and the correlation between pulse height and input photon number is completely lost.

In a real experiment, one uses a discriminator circuit to identify a pulse corresponding to a photon-detection event. The output photocurrent always carries noise from the electronic circuit. When the ENF is small (for example, $F = 1.05$), one can set the discriminator level between 0 and the distribution and tell the signal pulse from the background noise in the electronics without much ambiguity. Such discrimination becomes increasingly difficult as F increases, and when it reaches $F = 2$, it is impossible to tell the small pulses

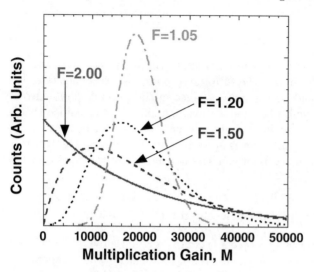

Fig. 12.7. Pulse height distribution calculated for various values of F

from the electronic background noise. When the discriminator level is set too high, the QE decreases, since the photons which produce shorter pulses are not counted. When the discriminator level is set too low in an attempt to increase the QE, the dark count increases since the electronic noise is now counted as pulses. Low multiplication noise below $F = 2$ is essential in order to avoid this compromise between QE and dark count.

12.7 Excess Noise Factor of a VLPC

In an APD, the impact-ionization excites an electron–hole pair across the entire bandgap. Such a device requires a large applied field so that the generated carriers can acquire enough kinetic energy to excite the next electron–hole pair. The presence of such a high field tends to make the multiplication process Markovian. Previous measurements on the pulse-height distribution of SSPMs and VLPCs [8, 237] show that it features a very well-defined peak with a narrow dispersion. These experimental results indicate that an ENF lower than two is realized in these devices.

Two mechanisms are proposed for low-noise performance of SSPMs and VLPCs. The electric field required for these devices is much smaller than that for an APD, since the gain mechanism is impact-ionization of electrons from neutral As atoms with energy levels only 54 meV below the conduction-band minimum. Single carrier-type multiplication is naturally realized, since the holes left in the impurity band relax through slow hopping conduction and do not acquire enough energy for further impact- ionization. Furthermore, due to the small electric field in the device, the generated electron has to be

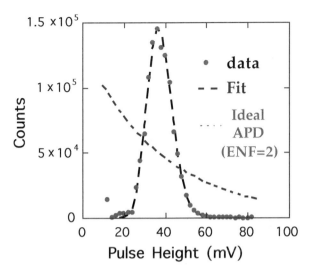

Fig. 12.8. A typical pulse-height distribution from the VLPC. The *dashed line* is the theoretical fit generated by using $F = 1.025$. *Dotted line* shows the theoretical pulse-height distribution expected for an ideal avalanche photodiode with $F = 2$

accelerated over a finite time period before it acquires enough kinetic energy for the next impact-ionization event. This "delay time" is expected to reduce the ENF in VLPCs [144]. Also, as the electrons are excited to the conduction band during avalanche multiplication, impurity holes are left behind, which relax on a time scale ($\sim$ μs) much longer than the multiplication process ($\sim$ ns). The positive space charge left behind accumulates as the multiplication proceeds; it modifies the local electric field, which tends to slow down the electrons and sustain the size of the avalanche [237]. These two effects lead to reduction in F below two towards the noise-free amplification of $F = 1$. In this section, results are presented for experiments in which the ENF for a VLPC is measured by two complementary methods.

12.7.1 Digital Measurement of the Pulse-Height Distribution

The first method is the analysis of the pulse-height distribution for electrical pulses generated from individual absorbed single photons and deduce the ENF, F, from their statistical mean and variance, as reported by several other authors [8, 237]. In this experiment, the signal pulse from the VLPC is amplified using a low-temperature GaAs FET amplifier. Due to the bandwidth limitation of the amplifier, the output pulse width is about 40 ns.

A typical result is shown in Fig. 12.8, where the pulse-height analysis was performed on a set of pulses each resulting from a single detected photon. The height distribution forms a peak with narrow dispersion, which is clearly separable from the background electrical noise. The variance in the pulse-height

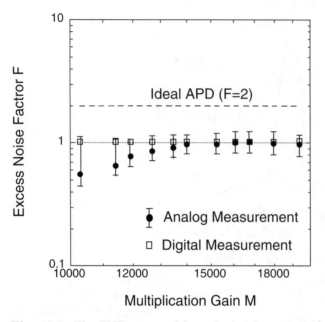

Fig. 12.9. The ENF measured by pulse-height analysis (*open squares*) and amplified shot noise measurement (*solid circles*). The *dotted line* is the ideal Si APD case, and $F = 1$ corresponds to noise-free avalanche multiplication

distribution comes from both the electrical readout circuit and fluctuation in the avalanche multiplication factor. The variance added by the electronics can be measured independently and subtracted to obtain the variance due to the internal multiplication mechanism. The theoretical fit (dashed line) was generated by using (12.6), using the mean and variance calculated from the data. For comparison, the pulse-height distribution expected for an ideal avalanche photodiode with $F = 2$ (exponential distribution) is shown as a dotted line. The analysis gives ENFs ranging between 1.02 and 1.04 and is plotted as open squares in Fig. 12.9. The average multiplication gain $\langle M \rangle$ can also be deduced from this measurement by carefully calibrating the gain and bandwidth of the amplifier circuit, since the pulse height is proportional to the total number of charges contained in each pulse.

12.7.2 Analog Noise Power Spectral Density Measurement

The second method to obtain the ENF is the measurement of the noise spectral density in the analog photocurrent resulting from an incident photon stream and the subsequent multiplication process, i.e., the noise power, $S_{\Delta I}$, is measured to estimate F from (12.5). Figure 12.10 shows the schematic of the noise measurement setup. Highly attenuated light-emitting diode (LED) light illuminates the VLPC, where the gain is controlled by the bias voltage,

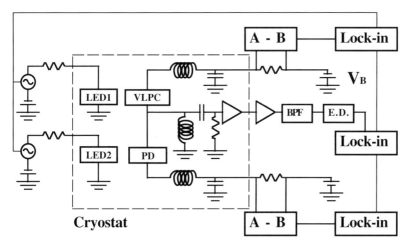

Fig. 12.10. Schematic of the experimental setup to measure the analog noise power spectral density from a VLPC. E.D.: envelope detector; PD: photodetector; BPF: band-pass filter

V_B. The noise signal is amplified by a low-temperature GaAs MESFET amplifier, which is cooled to 5~6 K with the VLPC. In order not to saturate the VLPC, one needs to work at a level of $\leq$10 fA of primary photocurrent. The generated noise power from such weak light signal is typically far less than 1×10^{-24} A^2/Hz (this corresponds to full shot noise due to ~1 µA of direct current), and several experimental techniques were adopted to improve the SNR ratio.

The GaAs MESFET amplifier has an input-equivalent voltage noise of about 2.6×10^{-18} V^2/Hz. In order to see the noise signal above this background thermal noise, one needs to use a load impedance of $\geq$10 kΩ. However, such a large load impedance will be easily shunted by the stray capacitance of the VLPC (~14 pF) at high frequencies, and thus the measurement bandwidth would be limited to below 1 MHz. In order to overcome this problem, a resonance circuit was used where the current generated by the VLPC flows through an inductive load [157, 209] which, with the capacitance of the VLPC and the parallel photodetector, forms a tank circuit of resonance frequency at 8.6 MHz. The effective load impedance increases by the quality factor $Q \simeq 35$ of the resonance circuit at 8.6 MHz, and large load impedance can be provided without sacrificing the measurement frequency.

The SNR ratio of the noise measurement setup can further be improved using a modulation technique [197]. The light incident on the VLPC was modulated at a low frequency (23 Hz), and the noise power was detected synchronously with the modulation frequency. Since the current is modulated, the noise power within the detection bandwidth ($\Delta f \simeq$ 220 kHz, determined by the resonance circuit) is also modulated. The signal is then amplified and fed into an envelope detector consisting of a high-frequency diode and a capacitor. This envelope detector outputs a voltage proportional to the incident noise power. The output of the envelope detector that corresponds to the light signal is modulated at 23 Hz on top of a DC voltage that corresponds to background noise. One can demodulate this AC component with a lock-in amplifier with a bandwidth of $\Delta\nu$. This measurement technique improves the SNR ratio by a factor of $\sqrt{\Delta f/\Delta\nu}$ [197], which in this case is slightly more than 10^3 using a lock-in time constant of 3 s. Combining the resonance circuit with the modulation technique, the measurement sensitivity was improved to $\sim$1.5 $\times 10^{-28}$ A^2/Hz (full shot noise corresponding to current levels of about 0.5 nA).

The measured noise power, P, in this experiment is given by

$$P = S_{\Delta I} Z_L^2 \Delta f G \quad , \tag{12.7}$$

where $S_{\Delta I}$ is the power spectral density in (12.5), Z_L is the load impedance, Δf is effective bandwidth and G is an effective gain of the amplifier and envelope detector circuit. It is very difficult to measure these quantities independently, so the product of these quantities were calibrated with a known noise source. A photodiode was inserted in parallel with the VLPC at the input of the amplifier. When the light incident on this photodiode is modulated, it generates a current carrying full shot noise at the input of the amplifier. By measuring this noise, P_{cal}, and the photocurrent, I_{cal}, through the photodiode, one can use the relation $P_{cal} = 2eI_{cal}Z_L^2\Delta f G \equiv A_0 I_{cal}$ to find the factor $A_0 = 2eZ_L^2\Delta f G$.

By measuring the current $\langle I_{ph}\rangle\langle M\rangle$ through the VLPC and the noise power P, one can deduce the product of the multiplication gain and the excess noise factor $\langle M\rangle$F. As explained above, the average multiplication gain $\langle M\rangle$ could be determined independently from pulse-height analysis, and thus F can be determined. The solid circles in Fig. 12.9 show the ENF determined from the analog noise power measurement. When pulses are sampled to measure $\langle M\rangle$, one has to set the trigger level to be able to tell pulses from background electrical noise. As the gain of the VLPC is reduced, the signal pulses get smaller and it becomes increasingly difficult to make the distinction. If one fixes the trigger level in the low-gain regime, one tends to preferentially sample taller pulses, thus overestimating the average gain. This accounts for the larger error bar on F on the upper side as the device gain is decreased. The measurement results show that the ENF for VLPCs is very close to unity, approaching noise-free avalanche multiplication. The measured values are clearly below the theoretical value of two expected for an ideal avalanche photodiode.

12.8 Two-Photon Detection with a VLPC

Besides the capability to unambiguously identify pulses from electronic noise, noise-free avalanche multiplication and the well-defined pulse-height distribution provide a strong correlation between the number of detected photons and the resulting pulse height. When two photons are detected, one can expect that the number of electrons released in the multiplication process is twice the typical value for a single-photon detection, resulting in a pulse that is twice as tall. Unlike an APD, where the whole diode breaks down upon detection of a photon, the avalanche breakdown in a VLPC is confined to a small portion ($\sim$20 μm diameter) of the total area (1 mm diameter) of the detector. This means that the remainder of the device is still active for another photon detection event even if a single photon is detected. These properties open up the possibility for multiple-photon detection using VLPCs.

Such behavior has already been observed in previous experiments where the time resolution of the electronic circuit was poor [8], so that more than two photons are detected by the VLPC as a single pulse. In these experiments, however, the difference in the arrival time of the two photons are reflected in the pulse height, which results in significant broadening of the pulse-height distribution. The exact QE of two-photon detection and the bit error rate for a two-photon detection event cannot be determined quantitatively from these experiments. In order to characterize the two-photon detection capability of a VLPC, one needs a source of two photons where the time interval between them can be controlled.

12.8.1 Twin Photon Generation in Optical Parametric Downconversion

Non linear interaction of light in a crystal can provide frequency conversion of photons. Some examples are second harmonic generation, and frequency upconversion and downconversion. Such interactions are used to generate a tunable source of coherent light starting from a laser in optical parametric amplifiers (OPAs) and oscillators (OPOs). When an optical crystal lacking inversion symmetry is pumped by a laser with frequency ω_p, there is a finite probability that this photon will break into two photons, with frequencies ω_s and ω_i, which satisfy the relation $\omega_p = \omega_s + \omega_i$. This process is called optical parametric downconversion [272]. The output port that carries photons with frequency ω_s is called the signal channel, and the output port that carries photons with frequency ω_i is called the idler channel. When the frequency of the photons in the signal and idler channels are the same ($\omega_s = \omega_i$), the situation is called degenerate optical parametric downconversion. The energies of the two photons are identical, and so twin photons are generated. Since a single photon is broken into two photons, the generation times of the two photons are strongly correlated. Since the momentum is conserved in the process, the momentam of the two photons generated are strongly correlated as well. When one considers a configuration where the signal and idler ports are focused onto the VLPC, the delay in the arrival times of the two photons can be controlled precisely by the optical-path-length difference.

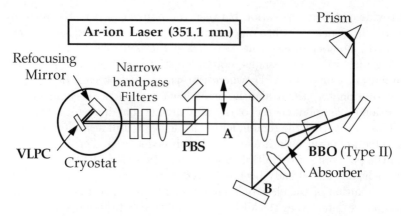

Fig. 12.11. Schematic setup for twin-photon generation using degenerate optical parametric downconversion and two-photon detection using a VLPC. PBS: polarizing beam splitter

Figure 12.11 shows the experimental setup for generating the twin photons. A beta barium borate (BBO) crystal is pumped by 351.1 nm ultraviolet radiation from an Ar$^+$ laser in a type-II phase-matching configuration. BBO is a negative uniaxial crystal, and in this phase-matching configuration the pump beam is polarized in the extraordinary direction. When the pump beam is tilted properly with respect to the optic axis, one can find two conical directions which satisfy the phase-matching condition for degenerate parametric downconversion. The polarization of photons in one cone is ordinary and in the other is extraordinary, so that these two photons have orthogonal polarization. The crystal was slightly tilted away from the collinear phase-matching condition (where the pump and the two output channels all have the same propagation direction), so that the signal and idler beams (both at 702.2 nm) were completely separated in space [137, 230]. Each of the beams was collimated using a weak focusing lens. One of the beams was delayed, and the two beams were recombined using a polarizing beam splitter. The recombined beam was focused onto the surface of the VLPC through a narrow bandpass filter (centered at 702 nm, and bandwidth of 0.26 nm full width at half-maximum [FWHM]). The VLPC was installed in a bath-type He cryostat where the temperature was stabilized to within 0.005 K using active temperature control. A similar shielding system for the room-temperature thermal radiation was used as that described in Sect. 12.4. The output signal from the VLPC was amplified by a room-temperature preamplifier (MITEQ AUX1347) which has a bandwidth of 500 MHz and provides electrical pulses of 2 ns in width when a single photon is detected.

It is very important to characterize the quality of the two-photon source in order to quantify the two-photon detection capability of the VLPC. The quality of the two-photon source is determined by the probability that a

Table 12.2. Count rates, dark-count rates and coincidence-count rates for the two-photon source using SPCM detectors. All numbers are given in units of counts per second (cps)

Quantity	Measured counts	Dark counts	Net Counts
A	1.22×10^5	2.59×10^3	2.38×10^5 (A_0)
B	8.41×10^4	1.37×10^4	1.41×10^5 (B_0)
C	1.13×10^4	negligible	4.53×10^4 (C_0)

photon is present in one output channel, provided that there is a photon in the other channel. This can be measured by placing single-photon counters at the two output channels, and looking for the events where both counters detect single photons simultaneously (coincidence counts). When the coincidence-count rate is equal to the count rate of each photon counter, it is a perfect two-photon source, since two photons are simultaneously detected with 100 % probability.

The narrow bandpass filter used in front of the VLPC had a transmittance of about 50 % at 702 nm, and such single-photon optical loss is responsible for the degradation of the two-photon source. The two-photon detection performance of the detection system itself can be estimated if such optical loss is subtracted. Two single-photon counting modules (SPCMs; single-photon counting detectors based on avalanche photodiodes) were used to characterize the optical loss and the quality of the two-photon source. Large area (500 μm diameter) APDs are employed in these SPCMs, and the QE of these detectors is ∼50 ±5 % near the measurement wavelength of 702 nm. The two SPCMs placed at locations A and B indicated in Fig. 12.11, with narrow bandpass filters in front of each detector, were used to measure the single-photon count rates of two beams (A and B, in cps) as well as the coincidence-count rate of the two beams (C). Given the dark counts of the SPCMs (which are measured by blocking the input beams) and the QE, one can estimate the net photon flux at the SPCM inputs (A_0 and B_0) and the net coincidence input (C_0) (Table 12.2). These three numbers characterize the single-photon loss rate of the two-photon source made up of a parametric downconverter and all the optical components used, including the narrow bandpass filters.

12.8.2 Characterization of Two-Photon Detection with VLPC

Figure 12.12a shows the real-time trace of an electrical pulse resulting from a single-photon detection event when one output beam of the parametric downconversion source is blocked. The width of the pulse (2 ns) does not decrease even when the bandwidth of the amplifier is increased, indicating that it is limited by the capacitance of the VLPC (∼ 14 pF) and the input impedance of the amplifier (50 Ω). Figure 12.12b and c show the cases when both beams are incident and the optical delay between the two beams is 5 ns

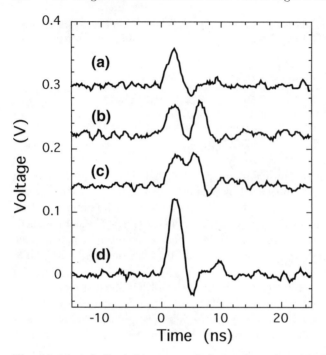

Fig. 12.12 a–d. Real-time trace of photon detection signal recorded by a 5 GS/s digitizing oscilloscope. The time delay between the two beams is changed by modifying the optical path length. The traces are shifted vertically for clarity. (a) Single-photon detection signal; (b) 5 ns delay; (c) 3 ns delay; (d) zero delay

and 3 ns, respectively. Two similar pulses are observed that are separated by the delay time. The heights of the pulses are almost identical, indicating that the number of electrons released per single-photon detection event is well defined. Finally, Fig. 12.12d shows when the optical delay between the two beams is reduced to zero. The two pulses resulting from the two-photon detection events completely overlap in time, and the pulse height is twice that of a single photon detection event.

Pulse-height analysis can be performed to estimate the bit error rate for the two-photon detection event. Figure 12.13 shows the pulse-height analyses of the cases when only one of the beams is incident (Fig. 12.13a) and when both beams are incident on the VLPC (Fig. 12.13b). For the two-beam incidence case, there is a second peak in the pulse-height distribution, centered at twice the value ($\sim$74 mV) of the center of the first peak ($\sim$37 mV).

From the pulse-height distribution obtained in the experiment, one can calculate the mean ($\langle M \rangle$) and ENF (F) of the gain, and use these values in (12.6) to generate the dotted lines of Fig. 12.13. For Fig. 12.13b, two curves were generated separately to fit each peak; they were then added to give the dotted line shown. The ENFs deduced from these pulse-height distribu-

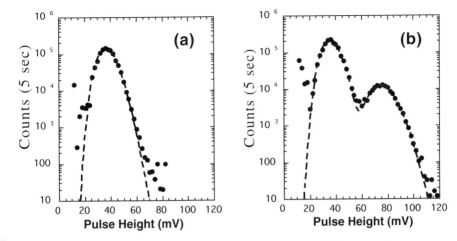

Fig. 12.13 a, b. Pulse-height distribution for the detected photons. Counts were integrated for 5 s at each point. The *solid circles* are experimental data, and the *dotted lines* are theoretical fits using Gamma distribution discussed in the text. (a) Single-beam input; (b) two-beam input with zero time delay between the two beams

tions were 1.026 for the single-photon pulse-height distribution (in both Fig. 12.13a and b) and 1.012 for the two-photon pulse-height distribution, indicating almost noise-free avalanche multiplication. The bit error rate P_e for distinguishing a two-photon detection event from a single-photon detection event is given by

$$P_e = \min_{V_T} \left\{ \int_{V_T}^{\infty} P_1(V)\, dV + \int_{-\infty}^{V_T} P_2(V)\, dV \right\} \quad , \tag{12.8}$$

where $P_1(V)$ and $P_2(V)$ are the normalized pulse-height distributions for single-photon detection events and two-photon detection events, respectively, and V_T is the threshold voltage used for the discrimination. From the two distributions given in Fig. 12.13b, P_e is minimized to 0.63 % when V_T is chosen at ∼54 mV. It should be noted that the discriminator level, V_T, for two-photon detection and the bit error rate, P_e, can change slightly depending on the relative size of the two peaks in Fig. 12.13b.

The ratio of the peaks is determined both by the quality of the two-photon source (dominated by the single-photon optical loss in the system) and the detector (QE of the two-photon detection event compared to single-photon detection event). Once the properties of the two-photon source are characterized (Table 12.2), one can estimate the two-photon detection efficiency of the VLPC. The ratio of the two-photon detection events and single-photon detection events is given by

$$\frac{R_2}{R_1} = \frac{\eta_2 C_0}{\eta_1 (A_0 + B_0 - 2\eta_1 C_0) + D_0} \quad , \tag{12.9}$$

where R_1 and R_2 denote the single-photon and two-photon count rates, η_1 and η_2 denote the single-photon and two-photon detection QE, and D_0 denotes the dark-count rate of the VLPC. Independent measurements at the same operating conditions yielded the values $\eta_1 = 70 \pm 5\%$ and $D_0 = 1.7 \times 10^4$ cps for the single-photon detection QE and the dark-count rate, respectively. Single-photon detection QE, η_1, was degraded by about 5% because of the saturation effect caused by the relatively large input photon flux ($\sim$5.6 $\times 10^5$ cps) used in the experiment [229]. The experimental value for R_2/R_1 can be found from integrating the pulse-height distribution curve in Fig. 12.13b; it was found to be 8.5×10^{-2}. Using these data with the values for A_0, B_0 and C_0 in Table 12.2, one can deduce the two-photon detection QE, $\eta_2 = 47\%$. The maximum value expected for a two-photon detection QE is given by $\eta_{2,\max} = \eta_1^2 = 49\%$, and the two-photon detection QE in a VLPC is limited by the single-photon detection QE within the measurement accuracy.

12.9 Summary

This chapter discussed the performance characteristics of a VLPC single-photon-counting system after comparing the merits and demerits of various single-photon detectors available at present. The operation principle of a VLPC was described, and the QE, dark counts and saturation characteristics of the VLPC were presented. The highest QE for single-photon detection ever reported (88%) was achieved with a VLPC in an optimally designed cryostat. The dependence of the QE and dark counts on temperature and bias voltage can be understood qualitatively from the operation principle of the VLPC. The intrinsic saturation characteristics of a VLPC can be explained quantitatively by the slow relaxation of impurity-band holes.

The physics behind the operation principle of a VLPC implies a low-noise avalanche multiplication process, which was also discussed in detail. Independent measurements show that the multiplication process in a VLPC features an ENF very close to one, approaching an ideal noise-free multiplication process. As a result of such a low-noise performance, the pulse-height distribution resulting from single-photon detection events feature a sharp peak with a very narrow dispersion. This opens up the possibility of two-photon detection using a VLPC. Experiments show that a photon detector based on VLPC can distinguish between single-photon incidence and two-photon incidence with high QE (47%), good time resolution (2 ns) and low bit error rate ($\sim$0.63 %). The performance of the detector was tested quantitatively using a two-photon source employing twin photons generated by a degenerate parametric downconversion process, where the optical delay between the two photons can be controlled precisely.

Such high quantum efficiency single photon detector can prove useful in applications such as quantum cryptography [35] and quantum teleportation [18, 28]. The two-photon detection capability can provide alternate means for test of Bell inequality [136, 184], and better realization of quantum teleportation [31, 29].

13. Future Prospects

13.1 Introduction

The radiative recombination of electron–hole pairs in a direct bandgap semiconductor is a fast and efficient process. Electron and/or hole injection through a p–n junction is a quiet process due to the collective- or single-charge Coulomb blockade effect. Combination of the above two mechanisms is at the heart of nonclassical light generation in a semiconductor light emitter. Generation of sub-Poisson light by a light-emitting diode (LED) [231], amplitude-squeezed light by a semiconductor laser [161] and regulated single photons by a single-photon turnstile device [125] have been demonstrated. The application and usefulness of such nonclassical light is limited to a low-loss system, in which random deletion of photons is negligible. Laser spectroscopy, interferometry, optical memory readouts and short-distance communication links promise such high-transmission characteristic and are the potential application areas of nonclassical light. It is interesting to explore whether a diode-laser-pumped solid-state laser may benefit from the sub-Poissonian pump photon statistics. Construction of such a squeezed high-power semiconductor laser as a pump diode laser is, in principle, possible.

A quantum key distribution system based on single photons [17] employs highly attenuated Poisson light with an average photon number per pulse $\langle n \rangle < 1$. The information rate decreases with decreasing $\langle n \rangle$ but the security degrades with increasing $\langle n \rangle$, because the probability of finding more two photons per pulse increases with $\langle n \rangle$ [215]. Regulated single photons from the turnstile device can increase the information rate and the security simultaneously. However, in order to use such a single photon source in a quantum key distribution system, a practical turnstile device operating at a higher temperature must be developed.

In this chapter, we will discuss some of future directions in nonclassical light generation by semiconductor devices.

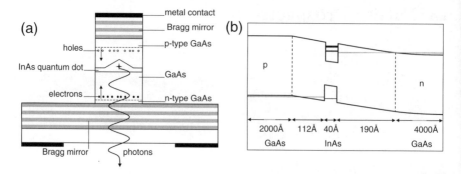

Fig. 13.1. (a) Proposed device structure. A single InAs QD (isolated on an etched micropost) is embedded in a GaAs p–i–n junction and surrounded by a microcavity. Metal contacts are made from the top and at the bottom substrate. Photons can be detected through the substrate. (b) Energy-band diagram of the structure. Doping levels are 10^{18} cm^{-3} on the n side and 10^{19} cm^{-3} on the p-side

13.2 Regulated and Entangled Photons from a Single Quantum Dot

Semiconductor quantum dots (QD's) [152] are very attractive for possible applications in electro-optic devices due to their atom-like properties and the strong confinement of electrons and holes [2, 138].

A fundamental nonlinear effect in a QD is the saturation of a single energy level by two electrons (or holes) of opposite spin due to the Pauli exclusion principle. We have proposed a device that produces regulated photons as well as pairs of entangled photons based on this effect [20].

Figure 13.1a illustrates a scheme for the device, which consists of a single InAs QD as the active medium embedded in a GaAs $p - j - n$ junction. Electrical contacts are made from a top metal contact and via the n^+ -doped substrate. The GaAs susbtrate is transparent for the ground state emission from the InAs QD's, and photons can be collected through the back side. It has been demonstrated that single QD's can be isolated from an ensemble of self-assembled QD's by etching small mesa or post structures [166], as sketched in the figure. The structure is surrounded by an optical microcavity, which modifies the spatial emission pattern and increases the spontaneous emission rate into resonant cavity modes [23]. A very large fraction β of spontaneous photons is thus emitted into a single mode of the cavity, and the outcoupling efficiency from the high refractive index material is improved as well. For the present state of the art, β values as high as 0.9 should be possible [73].

An energy band diagram of the structure is shown in Fig. 13.1b for doping levels of 10^{18} cm^{-3} on the n side and 10^{19} cm^{-3} on the p side. The QD layer is separated from the n (p) side by 190 Å (112 Å) wide GaAs intrinsic layers, which act as tunnel barriers. We assumed a typical dot diameter of 20 nm and height of 4 nm. For a qualitative discussion of the device operation, the Coulomb blockade energy can be estimated in a single particle picture [49, 58] for simplicity, with strain and piezoelectric effects [84] neglected. We assumed that the one and two electron ground state energy levels are 210 meV and 190 meV below the conduction band edge of GaAs, respectively, and that the one and two hole ground state energy levels are 100 meV and 80 meV above the valence band edge of GaAs, respectively. The first excited electron (p-like) state is about 70 meV above the ground state [64, 172]. These values are consistent with experimental observations [58, 64] and calculations [186, 253]. If the junction voltage V_j is well below the built-in potential, the carrier transport takes place by resonant tunneling of electrons and holes.

Figure 13.2 shows the calculated resonant tunneling rates for electrons and holes vs. the applied bias voltage. The calculation uses the WKB approximation with an effective electron and hole mass of $0.067\,m_0$ and $0.082\,m_0$, respectively, and a temperature of 4 K. The different lines correspond to the following (from left to right): Electron tunneling into the dot containing zero or one electron (solid lines) and hole tunneling into the dot containing two electrons and zero or one hole (dashed lines). Tunneling into the first excited electron state is indicated by dotted lines, where the three lines correspond to two electrons and two, one, or zero holes. The difference in the widths of electron and hole tunneling resonances is due to the asymmetric tunnel barriers and different doping levels. We chose the position of the QD within the GaAs layer in order to have the first hole resonant tunneling condition fulfilled at a junction voltage above the second electron tunneling resonance. In this situation we can switch on and off hole and electron tunneling by switching between different bias voltages.

Two-photon turnstile operation is achieved as follows: At a low bias voltage V_e (indicated in Fig. 13.2), two electrons can tunnel into the initially empty QD. Further electron tunneling is now completely suppressed due to the Pauli exclusion principle, since the ground state is filled and the next available electron state, the first excited state, is far off of resonance. Then, we switch up to a higher voltage V_h (indicated in Fig. 13.2), where two holes tunnel. Again, further hole tunneling is suppressed due to Pauli exclusion principle since the hole ground state is filled and the first excited hole state (not shown) is off resonance. The first excited electron state shifts by typically 7 mV [64] to lower voltages when a hole tunnels. This is indicated by the three dotted lines in Fig.13.2. However, even after two holes have tunneld into the QD, electron tunneling is inhibited. Once the holes have tunneled, radiative recombination annihilates two holes and produces exactly two photons. Thus, modulating the bias voltage between V_e and V_h produces a regulated stream of photons, where two photons are emitted per modulation cycle.

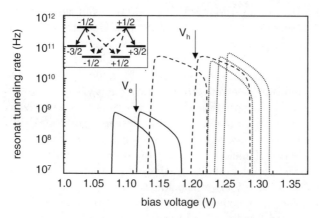

Fig. 13.2. Calculated resonant tunneling rates at 4 K into the QD ground state for electrons (*solid lines*) and holes (*dashed lines*) versus the applied bias voltage. Tunneling into the first excited electron state is indicated by dotted lines. In turnstile operation, the bias voltage is modulated between V_e and V_h. The inset illustrates optical transitions in a cubic lattice. The numbers indicate the projection of the total angular momentum J_z for the electrons and holes

The two photons arise from the decay of the biexcitonic ground state of the QD, where the correlated electrons and holes have opposite spins. If this anticorrelation translates into an anticorrelation in polarization of the emitted photons, it is easy to realize a single photon turnstile operation by selecting out only one photon per modulation cycle with the help of a polarizer. For quantum wells in direct-gap materials with a cubic lattice, any photons emitted are circularly polarized, because the $J_z = \pm 1/2$ electron recombines with the $J_z = \pm 3/2$ heavy hole [244]. This is illustrated in the inset in Fig. 13.2, where solid arrows indicate the σ^+ and σ^- ground state transitions. In the case of a QD, the strong confinement introduces level mixing and the hole ground state may have contributions from the $J_z = \pm 1/2$ hole states. Possible transitions to the $J_z = \pm 1/2$ states are indicated by dashed arrows in Fig. 13.2. Accordingly, when a $J_z = +1/2$ electron radiatively recombines with a hole in a QD, the emitted light is predominately σ^+ polarized, but may also have a σ^- component. Thus, the two photons that arise from the decay of the biexcitonic ground state are not necessarily perfectly anticorrelated with respect to σ^+ and σ^- polarization. An asymmetric dot shape, strain, and piezoelectric effects [217] further reduce the anticorrelation. However, there is experimental evidence from polarized photoluminescence [235] and two photon absorption measurements [32] that the anticorrelation in σ^+ and σ^- polarization is preserved in QD's. An exact calculation of the energy levels and oscillator strength including spin for the system discussed here would be desirable (so far optical and electronic properties of self-assembled InAs QD's have been calculated neglecting spin [217]).

We point out that a previous single photon turnstile device relies on the relatively small Coulomb splitting [125]. This limits the operation of this device to very low temperatures (40 mK) in order to guarantee that thermal energy fluctuations are negligible. In the proposed device, the turnstile operation is maintained up to much higher temperatures due to the very large splitting between the electron and hole ground and excited states. Electron and hole tunneling could be controlled merely by the Pauli exclusion principle, even if the Coulomb blockade effect were absent. For the parameters we assumed here, an operation at above 20 K should be possible. At higher temperatures, the electron and hole tunneling curves are broadened, mainly due to the thermal energy distribution of the electrons and holes in the $n-$ and $p-$doped layers. The broadening leads to a significant hole (electron) tunneling rate at lower (higher) bias voltage V_e (V_h), and photon emission can no longer be controlled. With a smaller QD and a larger splitting between ground and excited states, a larger broadening could be tolerated and thus a higher temperature operation is possible. We calculated that, up to a temperature of 50 K, thermionic emission can be neglected in the proposed structure.

We now focus on a very unique property of the proposed device, which is the production of pairs of entangled photons at well-defined time intervals. Starting from the biexcitonic ground state of the QD, a first electron can recombine with a hole and emit a σ^+ or a σ^- photon. Then, the second electron of opposite spin recombines with a hole, and a photon of opposite polarization is emitted. This situation is very similar to a two-photon cascade decay in an atom [7]. The two-photon state has the same form in any basis and is a maximally entangled (Bell) state: $|\psi\rangle = (|\sigma^+\rangle_1|\sigma^-\rangle_2 + |\sigma^-\rangle_1|\sigma^+\rangle_2)/\sqrt{2}$. Because of additional binding energy, the biexcitonic ground state has a smaller energy than twice the excitonic ground state [32]. Therefore, the first emitted photon 1 and the second emitted photon 2 have different energies (by approximately 4 meV).

The advantage of the proposed structure compared to other sources of entangled photons, such as two-photon cascade decay in atoms or parametric down-conversion in nonlinear crystals, is that entangled photon pairs are provided one by one with a tunable but regulated repetion rate of up to 1 GHz by a compact semiconductor device. The source is electrically pumped and the photons are emitted in resonant modes of an optical resonator, which greatly improves, e.g., the efficiency of subsequent fiber coupling. Such a regulated twin-photon source may be useful in a quantum key distribution system and other quantum network based on entangled photon-pairs.

A nearly perfect photon anti-bunching effect was in fact observed in a single InAs quantum dot which is isolated from a small etched post- structure [203] (see Fig. 1.4). In this experiment, optical pumping by a mode-locked Ti:Al$_2$O$_3$ laser was employed instead of electrical pumping via a pn junction. One of the advantages of a resonant optical pumping scheme is that high spectral purity of a pump laser allows the higher temperature operation of the device. Since the energy difference between the ground state 1e-1h transition and the excited state 2e-2h transition is much larger than $k_B T$ at a room temperature, a room temperature single photon source can be constructed.

13.3 Single-Mode Spontaneous Emission from a Single Quantum Dot in a Three-Dimensional Microcavity

Spontaneous emission of an atom is often thought of as an inherent property of the atom, but is in fact a result of the interaction between the atom dipole and the vacuum electromagnetic fields. Therefore, the radiation emitted from an atom can be altered by suitably modifying the surrounding vacuum fields with a cavity [57, 193]. This has been verified by pioneering experiments in the 1980's using atoms trapped in or passed through a cavity, which exhibit enhanced and inhibited spontaneous emission [78, 68, 99, 93]. Researchers are currently applying this principle, called cavity quantum electrodynamics (QED), to semiconductor nanostructures [195]; of particular interest is a single photon source for quantum cryptography. Here, we describe a semiconductor analog of a single atom cavity QED experiment, using a single quantum dot (QD) as the artificial atom, which is positioned in a three-dimensional post microcavity and coupled to a single optical mode. Because all the emission is from a single QD, the integrated spontaneous emission coupling efficiency into a single (polarization-degenerate) spatial mode of the device is nearly ideal, up to 78%.

The radiative transition rate of an atom from an excited, initial state to a lower energy, final state depends on the availability of photon field states. This is conveniently expressed by Fermi's golden rule as $(2\pi/\hbar)\rho(v_c)|\langle f|H|i\rangle|^2$, where $\rho(v_c)$ is the photon field density of states at the transition frequency, v_c, $|f>$ is the final state, $|i>$ is the initial state and H is the atom-vacuum field dipole interaction Hamiltonian. Thus, by modifying $\rho(v_c)$ using a periodic structure or a cavity, the spontaneous emission can be altered. In the weak-coupling regime, where the atomic excitation is irreversibly lost to the field, the spontaneous emission decay rate into a particular cavity mode is enhanced by increasing $\rho(v_c)$ at this cavity resonance. The enhancement factor is simply the ratio of the density of field states at the atomic frequency inside the cavity to that in free space. The cavity reduces the number of allowed modes, but increases the vacuum field intensity in these resonant modes. The result is the suppression of spontaneous emission outside the cavity resonance, and the enhancement of the spontaneous emission within the cavity resonance. For a localized atom with a negligible linewidth that is on resonance at the antinode of the standing wave, the enhancement factor is $3Q\lambda^3/4\pi^2 V$, where Q is the cavity quality factor, λ is the transition wavelength and V is the cavity mode volume.

In semiconductor systems, enhanced and inhibited spontaneous emission from quantum well (QW) excitons has been demonstrated in the weak coupling regime using a planar microcavity [264, 273]. The microcavity is formed by two distributed-Bragg reflectors (DBRs) separated by a $m\lambda/2$ cavity region, where λ is the wavelength of the light in the material and m is an integer. Because of one-dimensional confinement of photon fields and excitons, the spontaneous emission rate can be only slightly modified. The spontaneous emission coupling coefficient β, defined as the fraction of total spontaneous emission from the source that is captured into the fundamental cavity mode, is less than 1% in planar microcavity structures. The three dimensional photonic bandgap structure has been proposed to increase the spontaneous emission coupling efficiency, β, close to one [259]. Using lithographic processing or lateral oxidation techniques, three-dimensional post microcavities have shown enhanced spontaneous emission efficiency and could be used to achieve the same goal [182, 73]. If β is made close to one, semiconductor devices such as a thresholdless laser [23, 80], sub-Poissonian light emitting diode [271] and single-photon turnstile device [271] could be realized. β is related to the modified spontaneous emission decay rate as $\beta = (\gamma - \gamma_0 - \gamma_c)/\gamma$, where γ is the enhanced spontaneous emission decay rate into the fundamental spatial mode of the cavity, γ_0 is the spontaneous emission decay rate in a homogeneous matrix, and γ_c is the fractional spontaneous emission decay rate into the solid angle of the cavity mode in the limit that the mirror reflectivity, $R \to 0$. Since this solid angle is at most a few degrees in our case, $\gamma_c \ll \gamma_0$.

The QD and planar microcavity were fabricated monolithically using molecular- beam epitaxy (MBE). The nanometer-scale QD was formed from InAs in a GaAs matrix by a strained-enhanced islanding process, commonly called self-assembly [226], where the lattice mismatch strain energy drives the InAs deposited on the growth surface to island. The islanding increases the surface energy, but the total energy is reduced by the decreased lattice-mismatch strain energy. Electrons and holes are confined within the QD by the InAs and GaAs bandgap discontinuity, and their energies are quantized because the QD size is comparable to their de Broglie wavelengths. Because of spatial localization, an electron and hole form a three-dimensionally confined exciton which absorbs and emits light at discrete wavelengths, just as do atoms. The planar cavity is a wavelength-thick layer of GaAs with 29.5 AlAs/GaAs DBR bottom mirror pairs and 15 top mirror pairs. The structure was patterned into arrays of microposts using electron-beam lithography and anisotropic Cl_2/Ar ECR dry etching. A schematic of the device and a scanning-electron microscope (SEM) image of a particular post are shown in Fig. 13.3a and b [278]. The cross-sectional area of the active layer of this device is approximately 0.04 μm^2, which approximates the minimum possible mode volume $[(\lambda/n)^3]$ if we take into account the vertical spread of the field into the top and bottom DBR regions. In this device there are approximately 3 QDs in the post. The large refractive-index contrast between

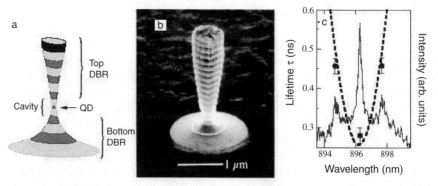

Fig. 13.3 a–c. A single quantum dot post microcavity. (a) A schematic of the device structure, which contains 15 top DBR pairs, a wavelength-thick (λ) cavity and 29.5 bottom DBR pairs. The DBR pairs are AlAs and GaAs, and the λ cavity contains a single InAs QD layer. The micropost is etched through the top DBR, the cavity, and a few lower DBR pairs. (b) An SEM image of a typical device, showing the tapered etch and contrasting AlAs and GaAs DBR layers. (c) A photoluminescence (PL) spectrum (solid) and the spectral dependence of the spontaneous emission lifetime (filled circles:experimental, dashed line:theory) at 4K of a tapered post microcavity with ≤ 0.5 µm cavity diameter. A single QD, resonant with the fundamental cavity mode, features a strong emission peak, while two QDs near the cavity mode edge produce weak emission lines. The spontaneous emission lifetime of the central QD is approximately 0.30 ns, which is 4.3 times shorter than the free-space spontaneous emission lifetime of QDs (approximately 1.3 ns)

the post and the vacuum provides transverse confinement and, in combination with the vertical confinement provided by the DBR cavity, leads to fully 3-dimensionally (3D) confined optical modes. [73, 196] In Fig. 13.3c, the photoluminescence (PL) spectrum from such a system is shown, where the emission from a single QD is in resonance with the fundamental cavity mode. The cavity resonance wavelength and quality factor, Q are determined under high pump power conditions, where the weak diffuse emission from the QD ensemble is filtered by the cavity. The independently measured modified spontaneous emission lifetime of the resonant QD is also shown, along with the theoretically predicted variation of lifetime with respect to the wavelength across the cavity mode. The theory is a first principles calculation, using no fitting parameters.

A planar microcavity has the cavity resonance centered at 932 nm with a cavity quality factor, Q of 2300. When microposts are formed, the lateral confinement increases the fundamental cavity mode resonance to higher energies (blue shift) and increases the mode separation. In Fig. 13.4, the fundamental cavity mode energy shift is plotted as a function of the effective post diameter, along with the theoretical results [278]. The insert in Fig. 13.4 is the PL spectrum of a 6 µm diameter microcavity post. The discrete modes

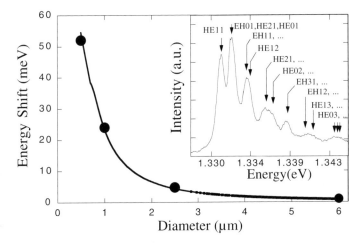

Fig. 13.4. Measured and calculated energy shifts of the fundamental cavity mode vs. post microcavity diameter. The experimental results are obtained from the PL spectral peak at 4K. The theoretical results are obtained by assuming the fields are separable into transverse and longitudinal components described, respectively, by dielectric waveguide modes and a transfer matrix formalism. In the insert, the PL spectrum at 4K from a 5 μm diameter post microcavity is shown together with the predicted cavity mode eigenenergies

are identified as the filtered spontaneous emission of the inhomogeneously broadened QDs. Quantitative agreement between theoretical and experimental resonance wavelengths is obtained without any free parameters.

Without a cavity, the QDs have a spontaneous emission lifetime of 1.3 ns, and with the planar microcavity this is reduced to 1.1 ns, as shown in Fig. 13.5a [278]. In the planar microcavity case, there is a continuous distribution of cavity resonant modes from the cut-off wavelength, $\lambda_c = 2d \simeq 932$ nm to a wavelength corresponding to the stopband edge. Since the number of modes increases with decreasing wavelength from the cutoff wavelength, λ_c, the spontaneous emission lifetime is continually decreased from the cutoff wavelength to shorter wavelengths. For the 3D post microcavities, the modes are discrete, so that the minimum spontaneous emission lifetime occurs at the center of each mode. This unmistakable mark of the 3D confinement is seen in Fig. 13.5b for the microcavity with a post diameter of 2.5 μm and also in Fig. 13.3c for a post diameter of 0.5 μm. For the 2.5 μm diameter post microcavity, a continuous distribution of QDs emit photons into the single cavity mode. The reduction of spontaneous emission lifetime is largest when the QDs are on resonance, but the modification disappears off-resonance, near the mode edge. In Fig. 13.5c the time-dependent spontaneous emission is shown for a single QD in resonance with a 0.5 μm diameter microcavity post ($\tau = 0.28$ ns), measured by a streak camera after the pulsed excitation from a 200 fs Ti:Al$_2$O$_3$ laser. The rise time (≈ 50 ps) is determined by the carrier

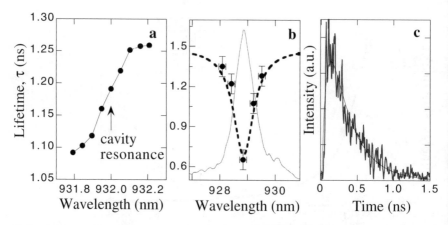

Fig. 13.5 a–c. Spontaneous emission lifetime and photoluminescence intensity vs. wavelength for (**a**) a QD planar microcavity and (**b**) a QD post microcavity with a diameter of 2.5 μm. In (**c**) the time-dependent spontaneous emission is shown for a single InAs QD, resonant with the fundamental mode of a 0.5 μm diameter post microcavity. The samples were pumped with a 200 fs, mode-locked Ti:sapphire laser tuned above the GaAs bandgap energy. Detection was made with a streak camera with 10 ps time resolution, with a spectrometer attachment so that the wavelength-dependent spontaneous emission lifetime is directly observed. The spontaneous emission rates were extracted as a function of wavelength by fitting an exponential rise and an exponential decay to the time-dependent intensity data, averaged over a spectral range of 0.25 nm

capturing time by the QD and the decay time (≈ 0.30 ns) is the radiative lifetime. The spontaneous emission lifetime at the center of each fundamental mode is plotted against the post diameter in the inset of Fig. 13.6, together with the theoretical prediction. The reduced spontaneous emission lifetime in a small post microcavity is primarily due to the reduced mode volume. This offsets the decrease in Q, which is approximately 300 in the 0.5 μm diameter post.

The spontaneous emission coupling coefficient is shown as a function of cavity post diameter in Fig. 13.6. Experimental values are extracted from the measured spontaneous emission rates using the formula given above. The solid line is the theoretical result with no free parameters. The effect of Q degradation in the smaller post structures is taken into account. For the smallest microcavity post, β is 0.78.

Other solid-state structures can be used for increasing β. Whispering gallery mode microdisks have realized a high cavity Q value [69], but isolation of a single QD in such structures has yet to be seen. Photonic bandgap structures have also been proposed for achieving a high spontaneous emission coupling coefficient, but this has also not yet been demonstrated. The single-mode spontaneous emission from a single QD suggests that nonclassical light based on such a single-mode spontaneous emission rather than stimulated

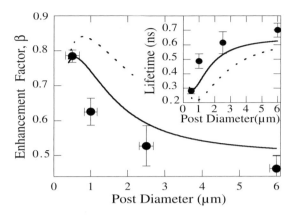

Fig. 13.6. Spontaneous emission coupling cofficient into the fundamental spatial mode, β, versus the post diameter. β for a single QD in the 0.5 μm diameter post is 0.78. Solid lines indicate the calculated values based on the enhancement factor formula given in the text, taking into account a random dipole orientation, the frequency resolution of the experiment and coupling to leaky modes. Degradation of the modal quality factor with decreasing post diameter is accounted for by calculating the diffraction loss into the lower DBR. In the insert, the experimental and theoretical spontaneous emission lifetimes are plotted as a function of the post diameter

emission in standard lasers is expected to play an important role in quantum communications based on single photons or photon twins.

Enhanced collection efficiency and decay rate of spontaneous emission from a quantum dot promise higher squeezing and a larger squeezing bandwidth if the pump process is regulated by either the Coulomb blockade effect or the Pauli exclusion principle [19].

13.4 Lasing and Squeezing of Exciton–Polaritons in a Semiconductor Microcavity

An exciton is a composite particle consisting of an electron and a hole. In a dilute limit, an exciton behaves like a quasi-bosonic particle. Due to its fermionic constituents, however, the exciton features strong exciton–exciton scattering, which provides enough stimulated emission gain for a ground-state exciton or microcavity exciton-polariton [232, 270]. Generation of coherent matter-wave (excitons and polaritons) by such an exciton laser (boser) has been demonstrated [98]. It is theoretically predicted that the exciton laser produces a number-phase squeezed state due to a strong self-phase modulation effect [233]. A semiconductor quantum-well microcavity generates not only nonclassical light [268] but also nonclassical de Broglie waves.

When moderate amounts of excitons are injected into a small microcavity post structure in which the exciton-polariton modes become quantized

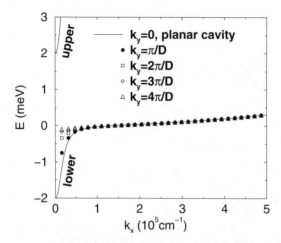

Fig. 13.7. The quantized energy/momenta for a 2 μm × 2 μm post structure. The lowest mode is the confined polariton mode, while other modes are densely spreaded on the energy axis, and can be considered as the exciton reservoir

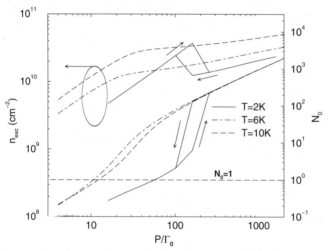

Fig. 13.8. Exciton density, left, and polariton population, right, for three different temperatures. $\tau_0 = 10ps$ and a 2×2 μm post structure was considered

(Fig. 13.7), large gain for the lowest confined polariton results from exciton-exciton scattering [232]. This gain is sufficient to compensate for typical losses for the exciton- polariton in GaAs based structures, and stimulated emission of the confined polaritons is eventually achieved (see Fig. 13.8). Under appropriate conditions, the threshold exciton density is significantly smaller than the critical density where the Rabi splitting of planar microcavities was found to collapse.

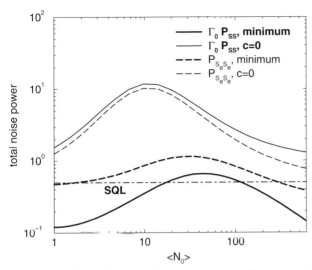

Fig. 13.9. Normalized number noise at zero frequency for the internal polariton S and external photon S_e as a functions of the polariton population $\bar{N}_0$, for T=6 K, τ_0=10 ps and a 2 × 2 μm post structure. c=0 corresponds to the bare number fluctuation and c=minimum correspond to the optimum compensation by the correlated frequency noise

Exciton-exciton interaction produces a self-phase modulation, resulting into enhanced frequency noise, and eventually, far above threshold, into a net increase of the emission linewidth. It is possible to use the correlation between frequency and number noise in order to produce a number-phase-squeezed state [233].

Figure 13.9 shows the normalized polariton number fluctuation with and without compensation by the correlated frequency noise vs. average polariton number at above the oscillation threshold. A large amount of squeezing is expected for the internal polariton number $\Gamma_o P_{SS}$ near the threshold and at far above threshold, if the number fluctuation is suppressed by referring to the frequency fluctuation (c = optimum value for minimizing P_{SS}).

Fig 10.3 ...

A. Appendix:
Noise and Correlation Spectra
for Light-Emitting Diode

A.1 Linearization

In this appendix, relations (2.26–2.28) are used to linearize (2.21–2.22). Linearizing gives

$$\frac{d\Delta N}{dt} = -\frac{\Delta N}{\tau_{sp}} + \frac{1}{\tau_{fi}}\frac{C_{dep}}{e}\Delta V_j - \frac{1}{\tau_{bi}}\Delta N - F_{sp} + F_{fi} - F_{bi}\,, \quad (A.1)$$

$$\frac{C_{dep}}{e}\frac{d\Delta V_j}{dt} = -\frac{\Delta V_j}{eR_s} - \frac{1}{\tau_{fi}}\frac{C_{dep}}{e}\Delta V_j + \frac{1}{\tau_{bi}}\Delta N - F_{fi} + F_{bi} + F_{rs}\,, \quad (A.2)$$

$$\Delta\Phi = \eta\frac{\Delta N}{\tau_{sp}} + \eta F_{sp} + F_v\,, \quad (A.3)$$

with time constants τ_{fi} and τ_{bi} defined by

$$\frac{1}{\tau_{fi}} = \frac{1}{C_{dep}}\frac{d}{dV_j}I_{fi}(V_j)\Big|_{V_j=V_0}\,, \quad (A.4)$$

$$\frac{1}{\tau_{bi}} = \frac{1}{e}\frac{d}{dN}I_{bi}(N)\Big|_{N=N_0}\,. \quad (A.5)$$

Taking the Fourier transform and solving for $\Delta\tilde{N}$, $\Delta\tilde{V_j}$ and $\Delta\tilde{\Phi}$ gives

$$\Delta\tilde{N} = \zeta(\Omega)\tau_{sp}\left[-\left(i\Omega\tau_{fi} + 1 + \frac{\tau_{fi}}{\tau_{RC}}\right)\tilde{F}_{sp}\right.$$

$$\left. -\left(i\Omega\tau_{fi} + \frac{\tau_{fi}}{\tau_{RC}}\right)(-\tilde{F}_{fi} + \tilde{F}_{bi}) + \tilde{F}_{rs}\right]\,, \quad (A.6)$$

$$C_{dep}\Delta\tilde{V_j} = \zeta(\Omega)\left[-\frac{\tau_{sp}\tau_{fi}}{\tau_{bi}}\vec{F}_{sp} + \tau_{fi}(1 + i\Omega\tau_{sp})(-\vec{F}_{fi} + \vec{F}_{bi})\right.$$

$$\left. +\tau_{fi}\left(1 + i\Omega\tau_{sp} + \frac{\tau_{sp}}{\tau_{bi}}\right)\tilde{F}_{rs}\right]\,, \quad (A.7)$$

$$\Delta\tilde{\Phi} = \eta\left[1 - \zeta(\Omega)\left(1 + i\Omega\tau_{fi} + \frac{\tau_{fi}}{\tau_{RC}}\right)\right]\tilde{F}_{sp}$$

$$-\eta\zeta(\Omega)\left(i\Omega\tau_{\text{fi}} + \frac{\tau_{\text{fi}}}{\tau_{\text{RC}}}\right)(-\vec{F}_{\text{fi}} + \vec{F}_{\text{bi}})$$
$$+\eta\zeta(\Omega)\vec{F}_{\text{rs}} + \vec{F}_{\text{v}}, \tag{A.8}$$

with $\tau_{\text{RC}} = R_{\text{s}}C_{\text{dep}}$ and

$$\zeta(\Omega) \equiv \left[1 - \Omega^2\tau_{\text{sp}}\tau_{\text{fi}} + \frac{\tau_{\text{fi}}}{\tau_{\text{RC}}}\left(1 + \frac{\tau_{\text{sp}}}{\tau_{\text{bi}}}\right)\right.$$
$$\left. + i\Omega\left(\tau_{\text{fi}} + \tau_{\text{sp}} + \frac{\tau_{\text{fi}}\tau_{\text{sp}}}{\tau_{\text{bi}}} + \frac{\tau_{\text{fi}}\tau_{\text{sp}}}{\tau_{\text{RC}}}\right)\right]^{-1}. \tag{A.9}$$

A.2 LED Photon Noise Spectral Density

The power spectral density of the output photon intensity fluctuation is calculated from (A.8), and can be expressed in terms of the noise power spectral densities of the individual noise sources as

$$S_{\Delta\Phi} = \eta^2\left|1 - \zeta(\Omega)\left(i\Omega\tau_{\text{fi}} + 1 + \frac{\tau_{\text{fi}}}{\tau_{\text{RC}}}\right)\right|^2 S_{F_{\text{sp}}} + \eta^2|\zeta(\Omega)|^2 S_{F_{\text{rs}}}$$
$$+\eta^2|\zeta(\Omega)|^2\left[\Omega^2\tau_{\text{fi}}^2 + \left(\frac{\tau_{\text{fi}}}{\tau_{\text{RC}}}\right)^2\right](S_{F_{\text{fi}}} + S_{F_{\text{bi}}}) + S_{F_{\text{v}}}. \tag{A.10}$$

The noise power spectral density of various noise sources can be calculated from their correlations defined in (2.14), (2.16), (2.18), (2.23) and (2.24):

$$S_{F_{\text{fi}}} = \frac{2I_{\text{fi}}(V_0)}{e}, \tag{A.11}$$

$$S_{F_{\text{bi}}} = \frac{2I_{\text{bi}}(N_0)}{e}, \tag{A.12}$$

$$S_{F_{\text{rs}}} = \frac{4k_{\text{B}}T}{e^2 R_{\text{s}}} = \frac{4(I + I_0)}{e}\frac{\tau_{\text{te}}}{\tau_{\text{RC}}}, \tag{A.13}$$

$$S_{F_{\text{sp}}} = \frac{2(I + 2I_0)}{e} = \frac{2I}{e} + \frac{4k_{\text{B}}T}{e^2 R_{\text{d0}}}, \tag{A.14}$$

$$S_{F_{\text{v}}} = 2\eta(1 - \eta)\frac{I + 2I_0}{e}, \quad 0 \le \eta \le 1. \tag{A.15}$$

The last equality of (A.13) follows from the definition of τ_{te} in (2.35). In (A.14), the reverse saturation current I_0 is replaced by $k_{\text{B}}T/eR_{\text{d0}}$, where R_{d0} is the differential resistance of the junction at zero bias. Using these expressions, one can derive the photon flux power spectrum

$$S_{\Delta\Phi} = \eta\frac{2I}{e}[1 - \eta\chi(\Omega)]$$
$$+\eta\frac{4k_BT}{e^2R_{d0}}\left\{1 - \eta\chi(\Omega)\left[1 + \frac{\tau_{te}}{\tau_{RC}}\left(1 + \frac{\tau_{fi}}{\tau_{RC}}\right) + \Omega^2\tau_{fi}\tau_{te}\right]\right\},$$
$$(A.16)$$

with

$$\chi(\Omega) \equiv |\zeta(\Omega)|^2$$
$$= \left\{\left[1 - \Omega^2\tau_{sp}\tau_{fi} + \frac{\tau_{fi}}{\tau_{RC}}\left(1 + \frac{\tau_{sp}}{\tau_{bi}}\right)\right]^2\right.$$
$$\left. +\Omega^2\left(\tau_{fi} + \tau_{sp} + \frac{\tau_{fi}\tau_{sp}}{\tau_{bi}} + \frac{\tau_{fi}\tau_{sp}}{\tau_{RC}}\right)^2\right\}^{-1}, \qquad (A.17)$$

where the relations $I_{fi}(V_0) = I + I_{bi}(N_0)$ and $I_{bi}(N_0) = (I + I_0)\tau_{sp}/\tau_{bi}$ were used.

A.3 External Current Noise Spectral Density

Since the external current is defined by $I_{ext} \equiv (V + V_s - V_j)/R_s$, its fluctuation is given by

$$\Delta I_{ext} = -\Delta V_j/R_s + eF_{rs}. \qquad (A.18)$$

From (A.7) and (2.32), one obtains the power spectral density of the external current fluctuation

$$S_{\Delta I_{ext}} = \chi(\Omega)\left[\left(\frac{\tau_{te}}{\tau_{RC}}\right)^2(2eI + 4k_BT/R_{d0})\right.$$
$$\left. + [1 + \Omega^2(\tau_{sp} + \tau_{te})^2]\frac{4k_BT}{R_s}\right], \qquad (A.19)$$

with $\chi(\Omega)$ defined by (A.17).

A.4 Junction-Voltage–Carrier-Number Correlation

When the bias is high $(I \gg I_0)$, $C_{n,v}(\Omega) \equiv N_{n,v}(\Omega)/G_n(\Omega)^{1/2}G_v(\Omega)^{1/2}$ can be calculated using (A.6), (A.7) and Eq. (2.32) as

$$N_{n,v}(\Omega) = \tau_{sp}\tau_{te}\left(1 + 2\frac{\tau_{te}}{\tau_{RC}}\right), \qquad (A.20)$$

$$G_n(\Omega) = \tau_{sp}^2\left(1 + 2\frac{\tau_{te}}{\tau_{RC}}\right), \qquad (A.21)$$

$$G_v(\Omega) = \tau_{te}^2\left(1 + 2\frac{\tau_{te}}{\tau_{RC}}\right). \qquad (A.22)$$

A.5 Photon-Flux–Junction-Voltage Correlation

One can calculate $C_{\Phi,v}(\Omega) \equiv N_{\Phi,v}(\Omega)/G_\Phi(\Omega)^{1/2}G_v(\Omega)^{1/2}$ for the high bias case $(I \gg I_0)$, using (A.7), (A.8) and (2.32). $G_v(\Omega)$ is given by (A.22) and

$$N_{\Phi,v}(\Omega) = \tau_{\text{te}} \left(\frac{\tau_{\text{te}}}{\tau_{\text{RC}}} + i\Omega(\tau_{\text{sp}} + \tau_{\text{te}}) \right), \qquad (\text{A.23})$$

$$G_\Phi(\Omega) = \frac{1}{\eta\chi(\Omega)} - 1. \qquad (\text{A.24})$$

References

1. J. Abe, G. Shinozaki, T. Hirano, T. Kuga, and M. Yamamishi, "Observation of the collective Coulomb blockade effect in a constant-current- driven high-speed light-emitting diode," *Journal of the Optical Society of America B*, vol. 14, pp 1295–1298, 1997.
2. F. Adler, M. Geiger, A. Bauknecht, F. Scholz, H. Schweizer, M. H. Pilkuhn, B. Ohnesorge, and A. Forchel, "Optical transitions and carrier relaxation in self assembled InAs/GaAs quantum dots," *Journal of Applied Physics*, vol. 80, pp 4019–4026, 1996.
3. G. P. Agrawal, "Gain nonlinearities in semiconductor lasers: theory and application to distributed feedback lasers," *IEEE Journal of Quantum Electronics*, vol. QE-23, pp 860–868, 1987.
4. M. H. Anderson, R. D. Jones, J. Cooper, S. J. Smith, D. S. Elliott, H. Ritsch, and P. Zoller, "Observation of population fluctuations in two-level atoms driven by a phase diffusing field," *Physical Review Letters*, vol. 64, pp 1346–1349, 1990.
5. T. Ando, M. Yasuda, and K. Sawada, "Noise suppression effect in an avalanche multiplication photodiode operating in a charge accumulation mode," *IEEE Transactions on Electron Devices*, vol. 42, pp 1769–1774, 1995.
6. A. Aspect, P. Grangier, and G. Roger, "Experimental tests of realistic local theories via Bell's theorem," *Physical Review Letters*, vol. 47, pp 460–463, 1981.
7. A. Aspect, J. Dalibard, and G. Roger, "Experimental test of Bell's inequalities using time-varying analyzers," *Physical Review Letters*, vol. 49, pp 1804–1807, 1982.
8. M. Atac, J. Park, D. Cline, D. Chrisman, M. Petroff, and E. Anderson, "Scintillating fiber tracking for the SSC using visible light photon counters," *Nuclear Instruments and Methods in Physics Research A*, vol. 314, pp 56–62, 1992.
9. M. Atac, "Visible light photon counter as high quantum efficiency photodetectors and applications," *Proceedings of the SPIE - The International Society for Optical Engineering*, vol. 2281, pp 54–64, 1994.
10. D. V. Averin and K. K. Likharev, "Coherent oscillations in small tunnel junctions," *Zhurnal Eksperimental'noi I Teoreticheskoi Fiziki*, vol. 90, pp 733–743, 1986 [Translated in: Soviet Physics – JETP].
11. D. V. Averin and K. K. Likharev, "Coulomb blockade of single-electron tunneling, and coherent oscillations in small tunnel junctions," *Journal of Low Temperature Physics*, vol. 62, 345–373, 1986.
12. J. B. Barner and S. T. Ruggiero, "Observation of the incremental charging of Ag particles by single electrons," *Physical Review Letters*, vol. 59, pp 807–810, 1987.
13. S. M. Barnett and C. R. Gilson, "Manipulating the vacuum: squeezed states of light," *European Journal of Physics*, vol. 9, pp 257–264, September 1988.

14. Th. Basché, W. E. Moerner, M. Orrit, and H. Talon, "Photon antibunching in the fluorescence of a single dye molecule trapped in a solid," *Physical Review Letters*, vol. 69, pp 1516–1519, 1992.

15. C. Becher, E. Gehrig, and K.-J. Boller, "Spectrally asymmetric mode correlation and intensity noise in pump-noise-suppressed laser diodes," *Physical Review A*, vol. 57, pp 3952–3960, 1998.

16. C. W. J. Beenakker and M. Büttiker, "Suppression of shot noise in metallic diffusive conductors," *Physical Review B*, vol. 46, pp 1889–1892, 1992.

17. C. H. Bennett, G. Brassard, S. Breidbart, and S. Wiesner, "Eavesdrop-detecting quantum communications channel," *IBM Technical Disclosure Bulletin*, vol. 26, pp 4363–4366, 1984.

18. C. H. Bennett, G. Brassard, C. Crepeau, R. Jozsa, A. Peres, and W. K. Wootters, "Teleporting an unknown quantum state via dual classical and Einstein–Podolsky–Rosen channels," *Physical Review Letters*, vol. 70, pp 1895–1899, 1993.

19. O. Benson and Y. Yamamoto, "Master equation model of a single quantum dot micorsphere laser," *Physical Review A*, vol. 59, pp 4756–4759, 1999.

20. O. Benson, C. Santori, M. Pelton, and Y. Yamamoto, "Regulated and entangled photons from a single quantum dot," *Physical Review Letters*, vol. 84, pp 2513–2516, 2000.

21. O. Benson, J. Kim, H. Kan, and Y. Yamamoto, "Simultaneous Coulomb blockade for electrons and holes in p-n junctions: Transition from Coulomb staircase to turnstile operation," *Physica E*, vol. 8, pp 5-12, 2000.

22. K. Bergman, C. R. Doerr, H. A. Haus, and M. Shirasaki, "Sub-shot-noise measurement with fiber-squeezed optical pulses," *Optics Letters*, vol. 18, pp 643–645, 1993.

23. G. Björk, S. Machida, Y. Yamamoto, and K. Igeta, "Modification of spontaneous emission rate in planar dielectric microcavity structures," *Physical Review A*, vol. 44, pp 669–681, 1991.

24. G. C. Bjorklund, "Frequency-modulation spectroscopy: a new method for measuring weak absorptions and dispersions," *Optics Letters*, vol. 5, pp 15–17, January 1980.

25. G. C. Bjorklund, M. D. Levenson, W. Lenth, and C. Oritz, "Frequency modulation (FM) spectroscopy: Theory of lineshapes and signal-to-noise analysis," *Applied Physics B*, vol. 32, pp 145–152, 1983.

26. R. S. Bondurant, P. Kumar, J. H. Shapiro, and M. M. Salour, "Photon-counting statistics of pulsed light sources," *Optics Letters*, vol. 7, pp 529–531, 1982.

27. M. Born and E. Wolf, *Principles of Optics: Electromagnetic Theory of Propagation, Interference and Diffraction of Light, 6th edn.* New York: Pergamon Press, pp 627–633, 1989.

28. D. Bouwmeester, P. Jian-Wei, K. Mattle, M. Eibl, H. Weinfurter, and A. Zeilinger, "Experimental quantum teleportation," *Nature*, vol. 390, pp 575–579, 1997.

29. D. Bouwmeester, J.-W. Pan, M. Daniell, H. Weinfurter, M. Zukowski, and A. Zeilinger, "Bouwmeester *et al.* reply," *Nature*, vol. 394, pp 841, 1998.

30. J. P. Bouyer, "Spectral stabilization of an InGaAsP semiconductor laser by injection locking," *Annales de Physique*, vol. 18, pp 89–239, April 1993.

31. S. L. Braunstein and H. J. Kimble, "*A posteriori* teleportation," *Nature*, vol. 394, pp 840–841, 1998.

32. K. Brunner, G. Abstreiter, G. Böhm, G, Träkle, and C. Weimann, "Sharp-line photoluminescence and two-photon absorption of zero- dimensional biexcitons

in a GaAs/AlGaAs structure," *Physical Review Letters*, vol. 73, pp 1138–1141, 1994.

33. M. J. Buckingham and E. A. Faulkner, "The theory of inherent noise in p-n junction diodes and bipolar transistors," *Radio and Electronic Engineer*, vol. 44, pp 125–140, 1974.

34. M. J. Buckingham, *Noise in Electronic Devices and Systems*. New York: Wiley, pp 71–95, 1983.

35. W. T. Buttler, R. J. Hughes, P. G. Kwiat, S. K. Lamoreaux, G. G. Luther, G. L. Morgan, J. E. Nordolt, C. G. Peterson, and C. M. Simmons, "Practical free-space quantum key distribution over 1 km," *Physical Review Letters*, vol. 81, pp 3283–3286, 1998.

36. H. Cao, G. Klimovitch, G. Björk, and Y. Yamamoto, "Theory of direct creation of quantum-well excitons by hole-assisted electron resonant tunneling," *Physical Review B*, vol. 52, 12184–12190, 1995.

37. F. Capasso, W. T. Tsang, and G. F. Williams, "Staircase solid-state photomultipliers and avalanche photodiodes with enhanced ionization rates ratio," *IEEE Transactions on Electron Devices*, vol. 30, pp 381–390, 1983.

38. F. Capasso, "Physics of avalanche photodiodes," in *Semiconductors and Semimetals* (eds. R. K. Willardson and A. C. Beer), *Lightwave Communications Technology* (ed. W. T. Tsang). New York: Academic Press, vol. 22, part D, pp 1–172, 1985.

39. C. M. Caves, K. S. Thorne, W. P. Drever, V. D. Sandberg, and M. Zimmermann, "On the measurement of a weak classical force coupled to a quantum-mechanical oscillator. I. Issues of principle," *Reviews of Modern Physics*, vol. 52, pp 341–392, 1980.

40. C. M. Caves, "Quantum-mechanical noise in an interferometer," *Physical Review D*, vol. 23, pp 1693–1708, 1981.

41. K. Chan, H. Ito, and H. Inaba, "Optical remote monitoring of CH4 gas using low-loss optical fiber link and InGaAsP light-emitting diode in 1.33-μm region," *Applied Physics Letters*, vol. 43, pp 634–636, 1983.

42. T. C. Chong, and C. G. Fonstad, "Theoretical gain of strained-layer semiconductor lasers in the large strain regime," *IEEE Journal of Quantum Electronics*, vol. QE-25, pp 171–178, 1989.

43. J. F. Clauser, M. A. Horne, A. Shimony, and R. A. Holt, "Proposed experiment to test local hidden-variable theories," *Physical Review Letters*, vol. 23, pp 880–884, 1969.

44. J. F. Clauser and A. Shimony, "Bell's theorem: experimental tests and implications." *Reports on Progress in Physics*, vol. 41, pp 1881–1927, 1978.

45. C. N. Cohen-Tannoudji and W. D. Phillips, "New mechanisms for laser cooling," *Physics Today*, vol. 43, pp 33–40, 1990.

46. J. A. Copeland, "Single-mode stabilization by traps in semiconductor lasers," *IEEE Journal of Quantum Electronics*, vol. QE-16, pp 721–727, 1980.

47. H. Dautet, P. Deschamps, B. Dion, A. D. MacGregor, D. MacSween, R. J. McIntyre, C. Trottier, and P. P. Webb, "Photon counting techniques with silicon avalanche photodiodes," *Applied Optics*, vol. 32, pp 3894–3900, 1993.

48. K. B. Dedushenko and A. N. Mamaev, "Phase and amplitude modulation in an injection-locked semiconductor laser," *Laser Physics*, vol. 3(5), pp 967–974, September-October 1993.

49. E. Dekel, D. Gershoni, E. Ehrenfreund, D. Spektor, J. M. Garcia, and P. M. Petroff, "Multiexciton Spectroscopy of a Single Self- Assembled Quantum Dot," *Physical Review Letters*, vol. 80, pp 4991–4994, 1998.

50. P. Delsing, K. K. Likharev, L. S. Kuzmin, and T. Claeson, "Time-correlated single-electron tunneling in one-dimensional arrays of ultrasmall tunnel junctions," *Physical Review Letters*, vol. 63, pp 1861– 1864, 1989.

51. F. De Martini, G. Di Giuseppe, and M. Marrocco, "Single-mode generation of quantum photon states by excited single molecules in a microcavity trap," *Physical Review Letters*, vol. 76, pp 900–903, 1996.

52. H. Deng and D. G. Deppe, "Oxide-confined vertical-cavity laser with additional etched void confinement," *Electronics Letters*, vol. 32, pp 900–901, 1996.

53. F. Diedrich and H. Walther, "Nonclassical radiation of a single stored ion," *Physical Review Letters*, vol. 58, pp 203–206, 1987.

54. K. P. Dinse, M. P. Winters and J. L. Hall, "Doppler-free optical multiplex spectroscopy with stochastic excitation," *Journal of the Optical Society of America B*, vol. 5, pp 1825–1831, 1988.

55. P. A. M. Dirac, *The Principle of Quantum Mechanics.* Oxford: Clarendon Press, 1958.

56. R. W. P. Drever, "Interferometric detectors for gravitational radiation," in: *Gravitational radiation, Les Houches 1982* (eds. N. Deruelle and T. Piran). Amsterdam: North-Holland, pp 321–338, 1983.

57. K. H. Drexhage, "Interaction of light with monomolecular dye layers," *Progress in Optics* (ed. E. Wolfe), North-Holland, Amsterdam, vol. 12, pp 165–232, 1974.

58. H. Drexler, D. Leonard, W. Hansen, J. P. Kotthaus, and P. M. Petroff, "Spectroscopy of quantum levels in charge-tunable InGaAs quantum dots," *Physical Review Letters*, vol. 73, pp 2252–2255, 1994.

59. P. H. Eberhard, "Background level and counter efficiencies required for a loophole-free Einstein–Podolsky–Rosen experiment," *Physical Review A*, vol. 47, pp R747–R750, 1993.

60. M. W. Fleming and A. Mooradian, "Spectral characteristics of external-cavity controlled semiconductor lasers," *IEEE Journal of Quantum Electronics*, vol. QE-17(1), pp 44–59, January 1981.

61. P. D. Floyd, B. J. Thibeault, L. A. Coldren, and J. L. Merz, "Scalable etched-pillar, AlAs-oxide defined vertical cavity lasers," *Electronics Letters*, vol. 32, pp 114–116, 1996.

62. M. J. Freeman, H. Wang, D. G. Steel, R. Craig, and D. R. Scifres. Amplitude-squeezed light from quantum-well lasers. *Optics Letters*, vol. 18(5), pp 379–381, March 1993.

63. M. J. Freeman, H. Wang, D. G. Steel, R. Craig, and D. R. Scifres. Wavelength-tunable amplitude-squeezed light from a room- temperature quantum-well laser. *Optics Letters*, vol. 18(24), pp 2141–2143, December 1993.

64. M. Fricke, A. Lorke, J. P. Kotthaus, G. Medeiros-Ribeiro, and P. M. Petroff, "Shell structure and electron-electron interaction in self- assembled InAs quantum dots," *Europhysics Letters*, vol. 36, pp 197–202, 1996.

65. P. Fritschel, D. Shoemaker, and R. Weiss, "Demonstration of light recycling in a Michelson interferometer with Fabry-Perot cavities," *Applied Optics*, vol. 31, pp 1412–1418, 1992.

66. T. A. Fulton and G. J. Dolan, "Observation of single- electron charging effects in small tunnel junctions," *Physical Review Letters*, vol. 59, pp 109–112, 1987.

67. A. Furusawa, J. L. Sorensen, S. L. Braunstein, C. A. Fuchs, H. J. Kimble, and E. S. Polzik, "Unconditional quantum teleportation," *Science*, vol. 282, pp 706–709, 1998.

68. G. Gabrielse and H. Dehmelt, "Observation of inhibited spontaneous emission," *Physical Review Letters*, vol. 55, pp 67–70, 1985.

69. B. Gayral, J. M. Gérard, A. LemaĜitre, C. Dupuis, L. Manin, and J. L. Pelouard, "High-Q wet-etched GaAs microdisks containing InAs quantum boxes," *Applied Physics Letters*, vol. 75, pp 1908–1910, 1999.

70. J. Gea-Banacloche and G. Leuchs, "Squeezed states for interferometric gravitational-wave detectors," *Journal of Modern Optics*, vol. 34, pp 793–811, 1987.

71. L. J. Geerligs, V. F. Anderegg, P. A. M. Howleg, J. E. Mooij, H. Pothier, D. Esteve, C. Urbina, and M. H. Devoret, "Frequency-locked turnstile device for single electrons," *Physical Review Letters*, vol. 64, pp 2691–2694, 1990.

72. N. Ph. Georgiades, R. J. Thompson, Q. Turchette, E. S. Polzik, and H. J. Kimble, "Spectroscopy with nonclassical light," *IQEC Conference Paper*, pp 222–223, 1994.

73. J. M. Gérard, B. Sermage, B. Gayral, B. Legrand, E. Costard, and V. Thierry-Mieg, "Enhanced spontaneous emission by quantum boxes in a monolithic optical microcavity," *Physical Review Letters*, vol. 81, pp 1110-1113, 1998.

74. K. E. Gibble, S. Kasapi, and S. Chu, "Improved magneto- optic trapping in a vapor cell," *Optics Letters*, vol. 17(7), pp 526-528, April 1992.

75. L. Gillner, G. Bjork, and Y. Yamamoto, "Quantum noise properties of an injection-locked laser oscillator with pump-noise suppression and squeezed injection," *Phyical Review A*, vol. 41, pp 5053–5065, 1990.

76. R. Glauber, "The quantum theory of optical coherence," *Physical Review*, vol. 130, pp 2529–2539 (1963); "Coherent and incoherent states of the radiation field," *ibid*, vol. 131, pp 2766–2788 (1963).

77. E. Goobar, A. Karlsson, and G. Björk, "Experimental realization of a semiconductor photon number amplifier and a quantum optical tap," *Physical Review Letters*, vol. 71, pp 2002–2005, 1993.

78. P. Goy, J. M. Raimond, M. Gross, and S. Haroche, "Observation of cavity-enhanced single-atom spontaneous emission," *Physical Review Letters*, vol. 50, pp 1903–1906, 1983.

79. H. Grabert and M. H. Devoret, *Single Charge Tunneling*, NATO ASI Series, vol. 294. New York: Plenum Press, 1992.

80. L. A.Graham, D. L. Huffaker, and D. G. Deppe, "Spontaneous lifetime control in a native-oxide-apertured microcavity," *Applied Physics Letters*, vol. 74, pp 2408-2410, 1999.

81. P. Grangier, R. E. Slusher, B. Yurke, and A. LaPorta, "Squeezed-light–enhanced polarization interferometer," *Physical Review Letters*, vol. 59, pp 2153–2156, 1987.

82. M. E. Greiner and J. F. Gibbons, "Diffusion and electrical properties of silicon-doped gallium arsenide," *Journal Applied Physics*, vol. 57, pp 5181–5187, 1985.

83. D. Grison, B. Lounis, C. Salomon, J. Y. Courtois, and G. Grynberg, "Raman spectroscopy of cesium atoms in a laser trap," *Europhysical Letters*, vol. 15(2), pp 149–154, May 1991.

84. M. Grundmann, O. Stier, and D. Bimberg, "InAs/GaAs pyramidal quantum dots: Strain distribution, optical phonons, and electronic structure," *Physical Review B*, vol. 52, pp 11969–11981, 1995.

85. N. Z. Hakim, B. E. A. Saleh, and M. C. Teich, "Generalized excess noise factor for avalanche photodiodes of arbitrary structure," *IEEE Transactions on Electron Devices*, vol. 37, pp 599–610, 1990.

86. J. L. Hall, L. Hollberg, T. Baer, and H. G. Robinson, "Optical heterodyne saturation spectroscopy," *Applied Physics Letters*, vol. 39, pp 680–682, 1981.

87. R. Hanbury Brown and R. Q. Twiss, "Correlation between photons in two coherent beams of light," *Nature*, vol. 177, pp 27–29, 1956.

88. W. Harth, W. Huber, and J. Heinen, "Frequency response of GaAlAs light-emitting diodes," *IEEE Transactions on Electron Devices*, vol. ED-23, pp 478–480, 1976.

89. H. A. Haus and Y. Yamamoto, "Quantum noise of an injection- locked laser oscillator," *Physical Review A*, vol. 29, pp 1261–1274, 1984.

90. E. R. Hegblom, B. J. Thibeault, R. L. Naone, and L. A. Coldren, "Vertical cavity lasers with tapered oxide apertures for low scattering loss," *Electronics Letters*, vol. 3, pp 869–871, 1997.

91. E. R. Hegblom, N. M. Margalit, A. Fiore, and L. A. Coldren, "Small efficient vertical cavity lasers with tapered oxide apertures," *Electronics Letters*, vol. 34, pp 895–896, 1998.

92. M. Heiblum, private communication.

93. D. H. Heinzen, J. J. Childs, J. E. Thomas, and M. S. Feld, "Enhanced and inhibited visible spontaneous emission by atoms in a confocal resonator," *Physical Review Letters*, vol. 58, pp 1320–1323, 1987.

94. M. Henny, S. Oberholzer, C. Strunk, T. Heinzel, K. Ensslin, M. Holland, and C. Schonenberger, "The fermionic Hanbury Brown and Twiss experiment," *Science*, vol. 284, pp 296–298, 1999.

95. C. H. Henry, "Theory of the linewidth of semiconductor lasers," *IEEE Journal of Quantum Electronics*, vol. QE-18, pp 259–264, 1982.

96. W. S. Hobson, F. Ren, U. Mohideen, R. E. Slusher, M. Lamont Schnoes, and S. J. Pearton, "Silicon nitride encapsulation of sulfide passivated GaAs/AlGaAs microdisk lasers," *Journal of Vacuum Science and Technology A*, vol. 13, pp 642–645, 1995.

97. C. K. Hong, S. R Friberg, and L. Mandel, "Optical communication channel based on coincident photon pairs," *Applied Optics*, vol. 24, pp 3877–3882, 1985.

98. R. Huang, F. Tassone, and Y. Yamamoto, "Experimental evidence of stimulated scattering of excitons into microcavity polariton," *Physical Review B*, vol. 61, pp R7854–R7857, March 2000.

99. R. G. Hulet, E. S. Hilfer, and D. Kleppner, "Inhibited spontaneous emission in a Rydberg atom," *Physical Review Letters*, vol. 55, pp 2137-2140, 1985.

100. K. Ikeda, S. Horiuchi, T. Tanaka, and W. Susaki, "Design parameters of frequency response on GaAs-(Ga,Al)As double heterostructure light- emitting diodes for optical communications," *IEEE Transactions on Electron Devices*, vol. ED-24, pp 1001–1005, 1977.

101. A. Imamoğlu, Y. Yamamoto, and P. Solomon, "Single- electron thermionic-emission oscillations in p-n microjunctions," *Physical Review B*, vol. 46, pp 9555–9563, 1992.

102. A. Imamoğlu and Y. Yamamoto, "Nonclassical light generation by Coulomb blockade of resonant tunneling," *Physical Review B*, vol. 46, pp 15982–15991, 1992.

103. A. Imamoğlu and Y. Yamamoto, "Noise suppression in semiconductor *p-i-n* junctions: Transition from macroscopic squeezing to mesoscopic Coulomb blockade of electron emission process," *Physical Review Letters*, vol. 70, pp 3327–3330, 1993.

104. A. Imamoğlu and Y. Yamamoto, "Turnstile device for heralded single photons: Coulomb blockade of electron and hole tunneling in quantum confined *p-i-n* heterojunctions," *Physical Review Letters*, vol. 72, pp 210–213, 1994.

105. S. Inoue, H. Ohzu, S. Machida, and Y. Yamamoto, "Quantum correlation between longitudinal-mode intensities in a multimode squeezed semiconductor laser," *Physical Review A*, vol. 46, pp 2757– 2765, 1992.

106. S. Inoue, S. Machida, Y. Yamamoto, and H. Ohzu., "Squeezing in an injection-locked semiconductor laser," *Physical Review A*, vol. 48, pp 2230–2234, 1993.

107. S. Inoue, G. Björk, and Y. Yamamoto, "Sub-shot-noise inteferometry with amplitude squeezed light from a semiconductor laser," *Proceedings of the SPIE*, vol. 2378, pp 99–106, 1995.

108. S. Inoue and Y. Yamamoto, "Longitudinal-mode-partition noise in a semiconductor-laser-based interferometer," *Optics Letters*, vol. 22, pp 328–330, 1997.

109. S. Inoue, S. Lathi, and Y. Yamamoto, "Longitudinal-mode- partition noise and amplitude squeezing in semiconductor lasers," *Journal of Optical Society of America B*, vol. 14, pp 2761–2766, 1997.

110. S. Inoue and Y. Yamamoto, "Gravitational wave detection using dual input Michelson interferometer," *Physical Letters*, vol. 236, pp 183–187, 1997.

111. R. Jackiw, "Minimum uncertainty product, number-phase uncertainty product, and coherent states," *Journal of Mathematical Physics*, vol. 9, pp 339–346, 1968.

112. P. P. Jenkins, A. N. MacInnes, M. Tabib-Azar, and A. R. Barron, "Gallium arsenide transistors: realization through a molecularly designed insulator," *Science*, vol. 263, pp 1751–1753, 1994.

113. J. B. Johnson, "Thermal agitation of electricity in conductors," *Nature*, vol. 119, pp 50–51, 1927.

114. A. Kapila, V. Malhotra, L. H. Camnitz, K. L. Seaward, and D. Mars, "Passivation of GaAs surfaces and AlGaAs/GaAs heterojunction bipolar transistors using sulfide solutions and SiN$_x$ overlayer," *Journal of Vacuum Science and Technology B*, vol. 13, pp 10–14, 1995.

115. S. Kasapi, S. Lathi, and Y. Yamamoto, "Sub-shot-noise FM spectroscopy with an amplitude squeezed semiconductor laser at room temperature." Talk presented at 7th Rochester Conference on Coherence and Quantum Optics (Rochester, NY, June 1995).

116. S. Kasapi, S. Lathi, and Y. Yamamoto, "Sub-shot-noise, frequency-modulated, diode-laser-based source for sub-shot-noise FM spectroscopy," *Optics Letters*, vol. 22, pp 478–480, 1997.

117. S. Kasapi, *Spectroscopic applications of amplitude- squeezed light from semiconductor lasers*, PhD Thesis, Department of Physics, Stanford University 1998.

118. S. Kasapi, S. Lathi, and Y. Yamamoto, "Sub-shot-noise FM noise spectroscopy of laser cooled rubidium atoms," *Journal of the Optical Society of America B*, vol. 15, pp 2626–2630, 1998.

119. P. L. Kelley and W. H. Kleiner, "Theory of electromagnetic field measurement and photoelectron counting," *Physical Review*, vol. 136, pp A316–A334,1964.

120. D. C. Kilper, M. J. Freeman, D. G. Steel, R. Craig, and D. R. Scifres, "Nonclassical amplitude squeezed light generated by semiconductor quantum well lasers," *Proceedings of the SPIE*, vol 2378, pp 64–76, 1995.

121. D. C. Kilper, A. C. Schaefer, J. Erland, and D. G. Steel, "Coherent nonlinear optical spectroscopy using photon-number squeezed light," *Physical Review A*, vol. 54(3), pp R1785–R1788, September 1996.

122. J. Kim, H. Kan, and Y. Yamamoto, "Macroscopic Coulomb-blockade effect in a constant-current-driven light-emitting diode," *Physical Review B*, vol. 52, pp 2008–2012, 1995.

123. J. Kim and Y. Yamamoto, "Theory of noise in p-n junction light emitters," *Physical Review B*, vol. 55, pp 9949–9959, 1997.

124. J. Kim, Y. Yamamoto, and H. H. Hogue, "Noise-free avalanche multiplication in Si solid state photomultipliers," *Applied Physics Letters*, vol. 70, pp 2852–2854, 1997.

125. J. Kim, O. Benson, H. Kan, and Y. Yamamoto, "A single photon turnstile device," *Nature*, vol. 397, pp 500–503, 1999.

126. J. Kim, S. Takeuchi, Y. Yamamoto, and H. H. Hogue, "Multi- photon detection using visible light photon counter," *Applied Physics Letters*, vol. 74, pp 902–904, 1999.

127. M. Kitagawa and Y. Yamamoto, "Number-phase minimum- uncertainty state with reduced number uncertainty in a Kerr nonlinear interferometer," *Physical Review A*, vol. 34, pp 3974–3988, 1986.

128. S. Kobayashi and T. Kimura, "Injection locking in AlGaAs semiconductor laser," *IEEE Journal of Quantum Electronics*, vol. QE-17, pp 681–689, 1981.

129. Kobayashi, "Direct frequency modulation in AlGaAs semiconductor lasers," *IEEE Journal of Quantum Electronics*, vol. QE-18, pp 582–595, 1982.

130. L. P. Kouwenhoven, A. T. Johnson, N. C. van der Vaart, C. J. P. M. Harmans, and C. T. Foxon, "Quantized current in a quantum- dot turnstile using oscillating tunnel barriers," *Physical Review Letters*, vol. 67, pp 1626–1629, 1991.

131. R. Kuchibhotla, J. C. Campbell, C. Tsai, W. T. Tsang, and F. S. Choa, "InP/InGaAsP/InGaAs SAGM avalanche photodiode with delta-doped multiplication region," *Electronics Letters*, vol. 27, pp 1361–1363, 1991.

132. A. Kumar, L. Saminadayar, D. C. Glattli, Y. Jin, and B. Etienne, "Experimental test of the quantum shot noise reduction theory," *Physical Review Letters*, vol. 76, pp 2778–2781, 1996.

133. P. G. Kwiat, A. M. Steinberg, R. Y. Chiao, P. H. Eberhard, and M. D. Petroff, "High-efficiency single-photon detectors," *Physical Review A*, vol. 48, pp R867–R870, 1993.

134. P. G. Kwiat, A. M. Steinberg, R. Y. Chiao, P. H. Eberhard, and M. D. Petroff, "Absolute efficiency and time-response measurement of single-photon detectors," *Applied Optics*, vol. 33, pp 1844–1853, 1994.

135. P. G. Kwiat, P. H. Eberhard, A. M. Steinberg, and R. Y. Chiao, "Proposal for a loophole-free Bell inequality experiment," *Physical Review A*, vol. 49, pp 3209–3220, 1994.

136. P. G. Kwiat, "Comment on 'Reliability of Bell-inequality measurements using polarization correlations in parametric-down-conversion photon sources'," *Physical Review A*, vol. 52, pp 3380–3381, 1995.

137. P. G. Kwiat, K. Mattle, H. Weinfurter, A. Zeilinger, A. V. Sergienko, and Y. Shih, "New high-intensity source of polarization-entangled photon pairs," *Physical Review Letters*, vol. 75, pp 4337–4341, 1995.

138. L. Landin, M. S. Miller, M.-E. Pistol, C. E. Pryor, and L. Samuelson, "Optical studies of individual InAs quantum dots in GaAs: few-particle effects," *Science*, vol. 280, pp 262–264, 1998.

139. D. V. Lang, R. L. Hartman, and N. E. Schmaker, "Capacitance spectroscopy studies of degraded $Al_xGa_{1-x}As$ DH stripe-geometry lasers," *Journal of Applied Physics*, vol. 47, pp 4986–4992, 1976.

140. R. Lang and K. Kobayashi, "External optical feedback effects on semiconductor injection laser properties," *IEEE Journal of Quantum Electronics*, vol. QE-16, pp 347–355, 1980.

141. S. Lathi, S. Kasapi, and Y. Yamamoto, "Phase-sensitive FM noise spectroscopy with diode laser," *Optics Letters*, vol. 21, pp 1000–1002, 1997.

142. S. Lathi and Y. Yamamoto, "Influence of nonlinear gain and loss on the intensity noise of multi-mode semiconductor lasers," *Physical Review A*, vol. 59, pp 819–825, 1999.

143. S. Lathi, K. Tanaka, T. Morita, H. Kan, and Y. Yamamoto, "Transverse-junction-stripe GaAs/AlGaAs lasers for squeezed light generation," *IEEE Journal of Quantum Electronics*, vol. 35, pp 387–394, 1999.

144. R. A. LaViolette and M. G. Stapelbroek, "A non- Markovian model of avalanche gain statistics for a solid-state photomultiplier," *Journal of Applied Physics*, vol. 65, pp 830–836, 1989.

145. A. T. Lee, "Broadband cryogenic preamplifiers incorporating GaAs MESFETs for use with low-temperature particle detectors," *Reviews of Scientific Instruments*, vol. 60, pp 3315–3322, 1989.

146. A. T. Lee, "A low-power-dissipation broadband cryogenic preamplifier utilizing GaAs MESFETs in parallel," *Reviews of Scientific Instruments*, vol. 64, pp 2373–2378, 1993.

147. T. P. Lee, "Effect of junction capacitance on the rise time of LED's and on the turn-on delay of injection lasers," *Bell System Technical Journal*, vol. 54, pp 53–68, 1975.

148. W. Lenth, C. Ortiz, and G. C. Bjorklund, "Pulsed frequency- modulation spectroscopy as a means for fast absorption measurements," *Optics Letters*, vol. 6, pp 351–353, 1981.

149. W. Lenth, "Optical heterodyne spectroscopy with frequency- and amplitude-modulated semiconductor lasers," *Optics Letters*, vol. 8, pp 575–577, 1983.

150. W. Lenth, "High frequency heterodyne spectroscopy with current-modulated diode lasers," *IEEE Journal of Quantum Electronics*, vol. QE-20, pp 1045–1050, 1984.

151. W. Lenth and G. Gehrtz, "Sensitive detection of NO_2 using high-frequency heterodyne spectroscopy with a GaAlAs diode laser," *Applied Physics Letters*, vol. 47, pp 1263–1265, 1985.

152. D. Leonard, M. Krishnamurthy, C. M. Reaves, S. P. Denbaars, and P. M. Petroff, "Direct formation of quantum-sized dots from uniform coherent islands of InGaAs on GaAs surfaces," *Applied Physics Letters*, vol. 63, pp 3203–3205, 1993.

153. M. D. Levenson, W. E. Moerner, and D. E. Horne, "FM spectroscopy detection of stimulated raman gain," *Optics Letters*, vol. 8, pp 108, 1983.

154. K. K. Likharev, "Correlated discrete transfer of single electrons in ultrasmall tunnel junctions," *IBM Journal of Research and Development*, vol. 32, pp 144–158, 1988.

155. D. L. Lile, "Interfacial constraints on III-V compound MIS devices," in: *Physics and Chemistry of III–V Compound Semiconductor Interfaces* (ed. C. W. Wilmsen). New York: Plenum Press, pp 327– 401, 1985.

156. R. C. Liu and Y. Yamamoto, "Suppression of quantum partition noise in mesoscopic electron branching circuits," *Physical Review B*, vol. 49, pp 10520–10532, 1994.

157. R. C. Liu, B. Odom, Y. Yamamoto, and S. Tarucha, "Quantum interference in electron collision," *Nature*, vol. 391, pp 263– 265, 1998.

158. R. C. Liu, *Quantum noise in mesoscopic electron transport*, PhD Thesis, Stanford University, 1997.

159. R. Loudon, *The Quantum Theory of Light*. New York: Oxford University Press, 1985.

160. W. H. Louisell, *Quantum Statistical Properties of Radiation*. New York: Wiley, 1973.

161. S. Machida, Y Yamamoto, and Y. Itaya, "Observation of amplitude squeezing in a constant-current-driven semiconductor laser," *Physical Review Letters*, vol. 58, pp 1000–1003, 1987.

162. S. Machida and Y. Yamamoto, "Ultrabroadband amplitude squeezing in a semiconductor laser," *Physical Review Letters*, vol. 60, pp 792–794, 1988.

163. S. Machida and Y. Yamamoto, "Observation of amplitude squeezing from semiconductor lasers by balanced direct detectors with a delay line," *Optics Letters*, vol. 14, pp 1045–1047, 1989.

164. L. Mandel and E. Wolf, *Optical Coherence and Quantum Optics*. Cambridge: Cambridge University Press, 1995.

165. F. Marin, A. Bramati, E. Giacobino, T. C. Zhang, J. Ph. Poizat, J. F. Roch, and P. Grangier, "Squeezing and intermode correlations in laser diodes," *Physical Review Letters*, vol. 75, pp 4606–4608, 1995.

166. J. Y. Marzin, J. M. Gérard, A. Izraël, D. Barrier, and G. Bastard, "Photo-luminescence of single InAs quantum dots obtained by self- organized growth on GaAs," *Physical Review Letters*, vol. 73, pp 716–719, 1994.

167. A. Massengale, M. C. Larson, C. Dai, and J. S. Harris Jr., "Collector-up Al-GaAs/GaAs heterojunction bipolar transistors using oxidised AlAs for current confinement," *Electronics Letters*, vol. 32, pp 399–401, 1996.

168. D. H. McIntyre, C. E. Fairchild, J. Cooper, and R. Walser, "Diode-laser noise spectroscopy of rubidium," *Optics Letters*, vol. 18, pp 1816–1818, 1993.

169. R. J. McIntyre, "Multiplication Noise in Uniform Avalanche Diodes," *IEEE Transactions on Electron Devices*, vol. ED-13, pp 164–168, 1966.

170. P. Mei, H. W. Yoon, T. Venkatesan, S. A. Schwarz, and J. P. Harbison, "Kinetics of silicon-induced mixing of AlAs-GaAs superlattices," *Applied Physics Letters*, vol. 50, pp 1823–1825, 1987.

171. J. L. Merz, J. P. Van der Ziel, and R. A. Logan, "Saturable optical absorption of the deep Te-complex center in $Al_{0.4}Ga_{0.6}As$," *Physical Review B*, vol. 20, pp 654-663, 1979.

172. B. T. Miller, W. Hansen, S. Manus, R. J. Luyken, A. Lorke, J. P. Kotthaus, S. Huant, G. Medeiros-Ribeiro, and P. M. Petroff, "Few-electron ground states of charge-tunable self-assembled quantum dots," *Physical Review B*, vol. 56, pp 6764–6769, 1997.

173. C. Monroe, W. Swann, H. Robinson, and C. Wieman, "Very cold trapped atoms in a vapor cell," *Physical Review Letters*, vol. 65, pp 1571–1574, 1990.

174. C. Monroe, D. M. Meekhof, B. E. King, W. M. Itano, and D. J. Wineland, "Demonstration of a fundamental quantum logic gate," *Physical Review Letters*, vol. 75, pp 4714–4717, 1995.

175. P. M. Mooney, G. A. Northrop, T. N. Morgan, and H. G. Grimneiss, "Evidence for large lattice relaxation at the DX center in Si-doped $Al_xGa_{1-x}As$," *Physical Review B*, vol. 37, pp 8298–8307, 1988.

176. A. Muller, J. Breguet, and N. Gisin, "Experimental demonstration of quantum cryptography using polarized photons in optical fibre over more than 1 km," *Europhysics Letters*, vol. 23, pp 383–388, 1993.

177. A. Muller, H. Zbinden, and N. Gisin, " Quantum cryptography over 23 km in installed under-lake telecom fibre," *Europhysics Letters*, vol. 33, pp 335–339, 1996.

178. R. S. Muller and T. I. Kamins, *Device Electronics for Integrated Circuits, 2nd edn.* New York: Wiley, pp 38–41, 1986.

179. S. Nakanishi, H. Ariki, H. Itoh, and K. Kondo, "Frequency-modulation spectroscopy of rubidium atoms with an AlGaAs diode laser," *Optics Letters*, vol. 12, pp 864–866, 1987.

180. H. Namizaki, H. Kan, M. Ishii, and A. Ito, "Transverse- junction-stripe-geometry double-heterostructure lasers with very low threshold current," *Journal of Applied Physics*, vol. 45, pp 2785–2786, 1974.

181. H. Nyquist, "Thermal agitation of electric charge in conductors," *Physical Review*, vol. 32, pp 110–113, 1928.

182. B. Ohnesorge, M. Bayer, A. Forchel, J. P. Reithmaier, N. A. Gippius, and S. G. Tikhodeev, "Enhancement of spontaneous emission rates by three-dimensional photon confinement in microcavities," *Physical Review B*, vol. 56, pp R4367–R4370, 1997.

183. W. D. Oliver, J. Kim, R. C. Liu, and Y. Yamamoto, "Hanbury Brown and Twiss-type experiment with electrons," *Science*, vol. 284, pp 299–301, 1999.

184. Z. Y. Ou and L. Mandel, "Violation of Bell's inequality and classical probability in a two-photon correlation experiment," *Physical Review Letters*, vol. 61, pp 50–53, 1988.

185. A. Peacock, P. Verhoeve, N. Rando, A. van Dordrecht, B. G. Taylor, C. Erd, M. A. C. Perryman, R. Venn, J. Howlett, D. J. Goldie, J. Lumley, and M. Wallis, "Single optical photon detection with a superconducting tunnel junction [for potential astronomical use]," *Nature*, vol. 381, pp 135–137, 1996.

186. F. M. Peeters and V. A. Schweigert, "Two-electron quantum disks," *Physical Review B*, vol. 53, pp 1468–1474, 1996.

187. D. T. Pegg and S. M. Burnett, "Phase properties of the quantized single-mode electromagnetic field," *Physical Review A*, vol. 39, pp 1665– 1675, 1989.

188. M. D. Petroff, M. G. Stapelbroek, and W. A. Kleinhans, "Detection of individual 0.4-μm wavelength photons via impurity-impact ionization in a solid-state photomultiplier," *Applied Physics Letters*, vol. 51, pp 406–408, 1987.

189. E. S. Polzik, J. Carri, and H. J. Kimble, "Spectroscopy with squeezed light," *Physical Review Letters*, vol. 68, pp 3020– 3023, 1992.

190. E. S. Polzik, J. Carri, and H. J. Kimble, "Atomic spectroscopy with squeezed light for sensitivity beyond the vacuum-state limit," *Applied Physics B*, vol. 55, pp 279–290, 1992.

191. H. Pothier, P. Lafarge, P. F. Orfila, C. Urbina, D. Esteve, and M. H. Devoret, "Single electron pump fabricated with ultrasmall normal tunnel junctions," *Physica B*, vol. 169, pp 573–574, 1991.

192. R. V. Pound, "Electronic frequency stabilization of microwave oscillators," *Reviews of Scientific Instruments*, vol. 17, pp 490–505, 1946.

193. E. M. Purcell, "Spontaneous emission probabilities at radio frequencies," *Physical Review*, vol. 69, pp 681, 1946.

194. E. L. Raab, M. Prentiss, A. Cable, S. Chu, and D. E. Pritchard, "Trapping of neutral sodium atoms with radiation pressure," *Physical Review Letters*, vol. 59, pp 2631–2634, 1987.

195. J. Rarity and C. Weisbuch, *Microcavities and Photonic Bandgaps: Physics and Applications*. Kluwer, Dordrecht, 1995.

196. J. P. Reithmaier, M. Rohner, H. Zull, F. Schafer, A. Forchel, P. A. Knipp, and T. L. Reinecke, "Size dependence of confined optical modes in photonic quantum dots," *Physical Review Letters*, vol. 78, pp 378–381, 1997.

197. M. Reznikov, M. Heiblum, H. Shtrikman, and D. Mahalu, "Temporal correlation of electrons: suppression of shot noise in a ballistic quantum point contact," *Physical Review Letters*, vol. 75, pp 3340–3343, 1995.

198. W. H. Richardson, S. Machida, and Y. Yamamoto, "Squeezed photon-number noise and sub-poissonian electrical partition noise in a semiconductor laser," *Physical Review Letters*, vol. 66, pp 2867–2870, 1991.

199. M. Rohner, J. P. Reithmaier, A. Forchel, F. Schafer, and H. Zull, "Laser emission from photonic dots," *Applied Physics Letters*, vol. 71, pp 488–490, 1997.

200. M. Romagnoli, M. D. Levenson, and G. C. Bjorklund, "Frequency-modulation-polarization spectroscopy," *Optics Letters*, vol. 8, pp 635–637, 1983.

201. C. J. Sandroff, R. N. Nottenburg, J.-C. Bischoff, and R. Bhat, "Dramatic enhancement in the gain of a GaAs/AlGaAs heterostructure bipolar transistor by surface chemical passivation," *Applied Physics Letters*, vol. 51, pp 33–35, 1987.

202. C. J. Sandroff, F. S. Turco-Sandroff, L. T. Florez, and J. P. Harbison, "Recombination at GaAs surfaces and GaAs/AlGaAs interfaces probed by *in situ* photoluminescence," *Journal of Applied Physics*, vol. 70, pp 3632–3635, 1991.

203. C. Santori, M. Pelton, G. Solomon, Y. Dale, and Y. Yamamoto, "Triggered single photons from a quantum dot," *Physical Review Letters*, vol. 86, pp 1502–1505, 2001.

204. E. Santos, "Unreliability of performed tests of Bell's inequality using parametric down-converted photons," *Physics Letters A*, vol. 212, pp 10–14, 1996.

205. M. Sargent III, M. O. Scully, and W. E. Lamb Jr., *Laser Physics*. Reading: Addison-Wesley Publishing Co., pp 327-340, 1974.

206. L. Schnupp, talk presented at European Collaboration Meeting on interferometric detection of gravitational waves, Sorrent, 1988.

207. S. Schroder, H. Grothe, and W. Harth, "Submilliampere operation of selectively oxidised GaAs-QW vertical cavity lasers emitting at 840 nm," *Electronics Letters*, vol. 32, pp 348–349, 1996.

208. B. L. Schumaker and C. M. Caves, "New formalism for two- photon quantum optics. II. Mathematical foundation and compact notation.," *Physical Review A*, vol. 13(5), pp 3093–3111, May 1985.

209. D. K. Serkland, M. M. Fejer, R. L. Byer, and Y. Yamamoto, "Squeezing in a quasi-phase-matched $LiNbO_3$ waveguide," *Optics Letters*, vol. 20, pp 1649–1651, 1995.

210. A. Shimizu and M. Ueda, "Effects of dephasing and dissipation on quantum noise in conductors," *Physical Review Letters*, vol. 69, pp 1403–1406, 1992.

211. J.-H. Shin, I.-Y. Han, and Y.-H. Lee, "Very small oxide-confined vertical microcavity lasers with high-contrast AlGaAs- Al_xO_y mirrors," *IEEE Photonics Technology Letters*, vol. 10, pp 754–756, 1998.

212. W. Shockley, "The theory of $p-n$ junctions in semiconductors and $p-n$ junction transistors," *Bell Systems Technical Journal*, vol. 28, pp 435–489, 1949.

213. D. Shoemaker, R. Schilling, L. Schnupp, W. Winkler, K. Maischberger, and A. Rudiger, "Noise behavior of the Garching 30-meter prototype gravitational-wave detector," *Physical Review D*, vol. 38, pp 423–432, 1988.

214. B. J. Skromme, C. J. Sandroff, E. Yablonovitch, and T. Gmitter, "Effects of passivating ionic films on the photoluminescence properties of GaAs," *Applied Physics Letters*, vol. 51, pp 2022–2024, 1987.

215. T. P. Spiller, "Quantum information processing: Cryptography, computation, and teleportation," *Proceedings of the IEEE*, vol. 84, pp 1719–1746, 1996.

216. M. G. Stapelbroek and M. D. Petroff, "Visible light photon counters for scintillating fiber applications: II. Principles of operation," in: *SciFi 93-Workshop on Scintillating Fiber Detectors* (eds. A. D. Bross, R. C. Ruchti, and M. R. Wayne). New Jersey: World Scientific, pp 621 629, 1993.

217. O. Stier, M. Grundmann, and D. Bimberg, "Electronic and optical properties of strained quantum dots modeled by 8-band k·p theory," *Physical Review B*, vol. 59, pp 5688–5701, 1999.

218. D. Stoler, "Equivalence classes of minimum uncertainty packets," *Physical Review D*, vol. 1, pp 3217-3219, 1970.

219. M. Sundaram, A. C. Gossard, and P. O. Holtz, "Modulation-doped graded structures: growth and characterization," *Journal of Applied Physics*, vol. 69, pp 2370–2375, 1991.

220. L. Susskind and J. Glogower, *Physics*, vol. 1, pp 49, 1964.

221. S. M. Sze, *Physics of Semiconductor Devices*. New York: Wiley, pp 21, 1981.

222. S. M. Sze, *Physics of Semiconductor Devices*. New York: Wiley, pp 56–57, 1981.

223. S. M. Sze, *Physics of Semiconductor Devices*. New York: Wiley, pp 74–84, 1981.

224. S. M. Sze, *Physics of Semiconductor Devices*. New York: Wiley, pp 84–96, 1981.

225. J. W. R. Tabosa, G. Chen, Z. Hu, R. B. Lee, and H. J. Kimble, "Nonlinear spectroscopy of cold atoms in a spontaneous-force optical trap," *Physical Review Letters*, vol. 66, pp 3245–3248, 1991.

226. M. Tabuchi, S. Noda, and A. Sasaki, "Mesoscopic structure in lattice-mismatched heteroepitaxial interface layers," in: *Science and technology of mesoscopic structures*, (eds. S. Namba, C. Hamaguchi, and T. Ando). Tokyo: Springer-Verlag, pp 379-384, 1992.

227. R. Takahashi, J. Mizuno, S. Miyoki, and N. Kawashima, "Control of a 10 m delay-line laser interferometer using the pre-modulation method," *Physics Letters A*, vol. 157–162, pp 157, 1994.

228. H. Takahasi, "Information theory of quantum-mechanical channels," in: *Advances in Communication Systems, vol. 1* (ed. A. V. Balakrishnan). New York: Academic Press, pp 227–310, 1965.

229. S. Takeuchi, J. Kim, Y. Yamamoto, and H. H. Hogue, "Development of high quantum efficiency single photon detection system," *Applied Physics Letters*, vol. 74, pp 1063–1065, 1999.

230. S. Takeuchi, talk presented at Meeting Abstracts of the Physical Society of Japan, vol. 53, pp 292, 1998.

231. P. R. Tapster, J. G. Rarity, and J. S. Satchell, "Generation of sub-Poissonian light by high-efficiency light-emitting diodes," *Europhysics Letters*, vol. 4, pp 293–299, 1987.

232. F. Tassone and Y. Yamamoto, "Exciton-exciton scattering dynamics in a semiconductor microcavity and stimulated scattering into polaritons," *Physical Review B*, vol. 59, pp 10830–10842, 1999.

233. F. Tassone and Y. Yamamoto, "Lasing and squeezing of composite bosons in a semiconductor microcavity," *Physical Review A*, vol. 62, pp 1–14, 2000.

234. M. C. Teich, K. Matsuo, and B. E. A. Saleh, "Excess noise factors for conventional and superlattice avalanche photodiodes and photomultiplier tubes," *IEEE Journal of Quantum Electronics*, vol. QE-22, pp 1184–1193, 1986.

235. Y. Toda, S. Shinomori, K. Suzuki, and Y. Arakawa, "Polarized photoluminescence spectroscopy of single self-assembled InAs quantum dots," *Physical Review B*, vol. 58, pp R10147–R10150, 1998.

236. Q. A. Turchette, C. J. Hood, W. Lange, H. Mabuchi, and H. J. Kimble, "Measurement of conditional phase shifts for quantum logic," *Physical Review Letters*, vol. 75, pp 4710–4713, 1995.

237. G. B. Turner, M. G. Stapelbroek, M. D. Petroff, E. W. Atkins, and H.H. Hogue, "Visible light photon counters for scintillating fiber applications: I. Characteristics and performance," in: *SciFi 93-Workshop on Scintillating Fiber Detectors* (eds. A. D. Bross, R. C. Ruchti, and M. R. Wayne). New Jersey: World Scientific, pp 613-620, 1993.

238. K. Vahala and A. Yariv, "Semiclassical theory of noise in semiconductor lasers. II," *IEEE Journal of Quantum Electronics*, vol. QE-19, pp 1102–1109, 1983.

239. A. Van der Ziel and E. R. Chenette, "Noise in solid state devices," *Advances in Electronics and Electron Physics*, vol. 46, pp 313–83, 1978.

240. J. Y. Vinet, B. Meers, C. N. Man, and A. Brillet, "Optimization of long-baseline optical interferometers for gravitational-wave detection," *Physical Review D*, vol. 38, pp 433–447, 1988.

241. D. F. Walls and G. J. Milburn, *Quantum Optics*. Berlin: Springer-Verlag, 1994.

242. H. Wang, M. J. Freeman, and D. G. Steel, "Squeezed light from injection-locked quantum well lasers," *Physical Review Letters*, vol. 71, pp 3951–3954, 1993.

243. Y. Wang, Y. Darici, and P. H. Holloway, "Surface passivation of GaAs with P_2S_5-containing solutions," *Journal of Applied Physics*, vol. 71, pp 2746–2756, 1992.

244. C. Weisbuch and B. Winter, *Quantum Semiconductor Structures*, San Diego: Academic Press, 1991.

245. E. A. Whittaker, C. M. Shum, H. Grebel, and H. Lotem, "Reduction of residual amplitude modulation in frequency-modulated spectroscopy by using harmonic frequency modulation," *Journal of the Optical Society of America B*, vol. 4(6), pp 1253–1256, June 1988.

246. C. E. Wieman and L. Hollberg, "Using diode lasers for atomic physics," *Review of Scientific Instruments*, vol. 62(1), pp 1–20, 1991.

247. G. F. Williams, F. Capasso, and W. T. Tsang, "The graded bandgap multilayer avalache photodiode: A new low-noise detector," *IEEE Electron Device Letters*, vol. 3, pp 71–73, 1982.

248. R. Williams, *Modern GaAs Processing Methods*. Boston: Artech House Inc., pp 133–137, 1990.

249. R. Williams, *Modern GaAs Processing Methods*. Boston: Artech House Inc., pp 140–146, 1990.

250. R. Williams, *Modern GaAs Processing Methods*. Boston: Artech House Inc., pp 153–187, 1990.

251. R. Williams, *Modern GaAs Processing Methods*. Boston: Artech House Inc., pp 211–227, 1990.

252. A. P. Willis, A. I. Ferguson, and D. M. Kane, "Longitudinal mode noise conversion by atomic vapour," *Optics Communications*, vol. 122, pp 32–34, 1995.

253. A. Wojs and P. Hawrylak, "Charging and infrared spectroscopy of self-assembled quantum dots in a magnetic field," *Physical Review B*, vol. 53, pp 10841–10845, 1996.

254. N. C. Wong and J. L. Hall, "Servo control of amplitude modulation in frequency-modulation spectroscopy: demonstration of shot-noise- limited detection," *Journal of the Optical Society of America B*, vol. 2, pp 1527–1533, 1985.

255. N. G. Woodard, E. G. Hufstedler, and G. P. Lafyatis, "Photon counting using a large area avalanche photodiode cooled to 100 K," *Applied Physics Letters*, vol. 64, pp 1177–1179, 1994.

256. M. Xiao, L.-A. Wu, and H. J. Kimble, "Precision measurement beyond the shot-noise limit," *Physical Review Letters*, vol. 59, pp 278- 281, 1987.

257. M. Xiao, L.-A. Wu, and H. J. Kimble, "Detection of amplitude modulation with squeezed light for sensitivity beyond the shot-noise limit, *Optics Letters*, vol. 13, pp 476–478, 1988.

258. E. Yablonovitch, C. J. Sandroff, R. Bhat, and T. Gmitter, "Nearly ideal electronic properties of sulfide coated GaAs surfaces," *Applied Physics Letters*, vol. 51, pp 439–441, 1987.

259. E. Yablonovitch,"Inhibited spontaneous emission in solid-state physics and electronics," *Physical Review Letters*, vol. 58, pp 2059–2062, 1987.

260. T. Yabuzaki, T. Mitsui, and U. Tanaka, "New type of high- resolution spectroscopy with a diode laser," *Physical Review Letters*, vol. 67, pp 2453–2456, 1991.

261. Y. Yamamoto, S. Saito, and T. Mukai, "AM and FM quantum noise in semiconductor lasers. II. Comparison of theoretical and experimental results for AlGaAs lasers," *IEEE Journal of Quantum Electronics*, vol. QE-19, pp 47–58, 1983.

262. Y. Yamamoto, S. Machida, and O. Nilsson, "Amplitude squeezing in a pump-noise-suppressed laser oscillator," *Physical Review A*, vol. 34, pp 4025–4042, 1986.

263. Y. Yamamoto and S. Machida, "High-impedance suppression of pump fluctuation and amplitude squeezing in semiconductor lasers," *Physical Review A*, vol. 35, pp 5114–5130, 1987.

264. Y. Yamamoto, S. Machida, K. Igeta, and Y. Horikoshi, "Enhanced and inhibited spontaneous emission of free excitons in GaAs quantum wells in a microcavity, in: *Coherence and Quantum Optics* (eds. J. H. Eberly, et al.). New York: Plenum Press, vol. 6, pp 1249–1257, 1989.

265. Y. Yamamoto, "Photon number squeezed states in semiconductor lasers," *Science*, vol. 255, pp 1219–1224, 1992.

266. Y. Yamamoto and H. A. Haus, "Effect of electrical partition noise on squeezing in semiconductor lasers," *Physical Review A*, vol. 45, pp 6596–6604, 1992.

267. Y. Yamamoto, "The quantum optical repeater," *Science*, vol. 263, pp 1394–1395, 1994.

268. Y. Yamamoto, "Squeezing and cavity QED in semiconductors," *Quantum Optics of Confined Systems* (eds. M. Ducloy and D. Bloch). Dordrecht: Kluwer, pp 201–281, 1996.

269. Y. Yamamoto, "A photon in solitary confinement," *Nature*, vol 390, pp 17–18, 1997.

270. Y. Yamamoto, "Semiconductor physics: Half-matter, half-light amplifier," *Nature*, vol. 405, pp 629-630, June 2000.

271. M. Yamanishi, K. Watanabe, N. Jikutani, and M. Ueda, "Sub-Poissonian photon-state generation by Stark-effect blockade of emissions in a semiconductor diode driven by a constant-voltage source," *Physical Review Letters*, vol. 76, pp 3432–3435, 1996.

272. A. Yariv, *Quantum Electronics, 3rd edn.* New York: John Wiley and Sons, Inc., pp 430–435, 1989.

273. H. Yokoyama, K. Nishi, T. Anan, H. Yamada, S. D. Brorson, and E. P. Ippen, "Enhanced spontaneous emission from GaAs quantum wells in monolithic microcavities," *Applied Physics Letters*, vol. 57, pp 2814– 2816, 1990.

274. H. P. Yuen, "Two-photon coherent states of the radiation field," *Physical Review A*, vol. 13, pp 2226-2243, 1976.

275. B. Yurke and E. A. Whittaker, "Squeezed-state-enhanced frequency-modulation spectroscopy," *Optics Letters*, vol. 12, pp 236– 238, 1987.

276. T. C. Zhang, J. Ph. Poizat, P. Grelut, J. F. Roch, P. Grangier, F. Marin, A. Bramati, V. Jost, M. D. Levenson, and E. Giacobino, "Quantum noise of free-running and externally-stabilized laser diodes," *Quantum and Semiclassical Optics: Journal of the European Optical Society Part B*, vol. 7, pp 601–613, 1995.

277. T. Zhang, P. J. Edwards, W. N. Cheung, and H. B. Sun, "Electrical characterisation of a quantum noise suppressed light emitting diode," *Technical Digest* (CLEO/ Pacific Rim, July 10-14, 1995, Chiba, Japan), pp 71–72, 1995.

278. G. Solomon, M. Pelton, and Y. Yamamoto, "Single-mode spontaneous emission from a single quantum dot in a 3D microcavity," Physical Review Letters, vol. 86, pp 3903–3906, 2001.

Subject Index

Druck: Strauss Offsetdruck, Mörlenbach
Verarbeitung: Schäffer, Grünstadt